METHODS IN MOLECULAR BIOLOGY™

Series Editor
John M. Walker
School of Life Sciences
University of Hertfordshire
Hatfield, Hertfordshire, AL10 9AB, UK

For other titles published in this series, go to
www.springer.com/series/7651

Biomedical Informatics

Edited by

Vadim Astakhov

Biomedical Informatics Research Network Coordination Center, The Center for Research in Biological Systems (CRBS), UCSD, La Jolla, CA

 Humana Press

Editor
Vadim Astakhov
Biomedical Informatics
   Research Network Coordination Center
The Center for Research
   in Biological Systems (CRBS)
UCSD, La Jolla
CA

ISSN 1064-3745   e-ISSN 1940-6029
ISBN 978-1-934115-63-3   e-ISBN 978-1-59745-524-4
DOI 10.1007/978-1-59745-524-4

Library of Congress Control Number: 2009926171

Printed on acid-free paper

Springer is part of Springer Science+Business Media (www.springer.com)

Preface

In recent decades, bioinformatics has emerged as a dynamic area producing a wide spectrum of new approaches and playing an important role in modern biotechnological development. This book provides an overview of novel cyberinfrastructures which are currently under development in various biomedical centers around the world. The first three chapters demonstrate various architectures for large-scale collaboration which integrate scientific accord across multiple centers. The next five chapters demonstrate modern approaches currently used in various areas of bioinformatics. The final four chapters illustrate the software challenges and strategies to resolve those challenges for large-scale biomedical informatics projects.

Unlike many other research areas which use standard techniques, development of biomedical cyberinfrastructure requires considerable expertise. The knowledge compiled in this book is contributed by experts in their fields who have many years of experience not only in using well-known techniques but also in development and troubleshooting them. This book might be helpful for working specialists as well as for students taking courses in bioinformatics and computer science. The chapters are arranged in a logical order, progressing from an overview of general architecture to detailed aspects of bioinformatics research. Consequently, each chapter contains a step-by-step description of the methodology and provides invaluable tips and tricks to safeguard against potential mishaps and pitfalls. It is a book that will be an essential manual for newcomers to this area, as well as a valuable addition to the laboratories and offices of more experienced researchers.

I would like to express my gratitude to all the authors who have made this book possible, and thank Professor John Walker, the Series Editor, for the invitation to edit this volume. My thanks also go to Ms. Meghan Kennelly for her assistance during the final stages of editing.

La Jolla, CA   *Vadim Astakhov*

Contents

Contributors

VADIM ASTAKHOV • *Biomedical Informatics Research Network Coordination Center, The Center for Research in Biological Systems (CRBS), UCSD, La Jolla, CA, USA*

TAMARA ASTAKHOVA • *Joint Center for Structural Genomics, University of California – San Diego, La Jolla, CA, USA*

MICHAEL BAITALUK • *San Diego Supercomputer Center, University of California – San Diego, La Jolla, CA, USA*

GREGORY G. BROWN • *Department of Psychiatry, University of California – San Diego, La Jolla, CA, USA*

CHRISTINE FENNEMA-NOTESTINE • *Departments of Psychiatry and Radiology, University of California – San Diego, La Jolla, CA, USA*

JEFFREY S. GRETHE • *Biomedical Informatics Research Network, La Jolla, CA, USA*

AMARNATH GUPTA • *San Diego Supercomputer Center, University of California – San Diego, La Jolla, CA, USA*

GYE WON HAN • *Joint Center for Structural Genomics (JCSG), The Scripps Research Institute, La Jolla, CA, USA*

LUKASZ JAROSZEWSKI • *The Burnham Institute, La Jolla, CA, USA*

DAVID B. KEATOR • *Brain Imaging Center, University of California – Irvine, Irvine, CA, USA*

DAVID R. LITTLE • *The Center for Research in Biological Systems (CRBS), La Jolla, CA, USA*

RAMIL V. MANANSALA • *Center for Research in Biological Systems, University of California – San Diego, La Jolla, CA, USA*

STEPHEN MONTGOMERY • *Wellcome Trust Sanger Institute, Hinxton, Cambridge, UK*

I. BURAK ÖZYURT • *Department of Psychiatry, University of California – San Diego, La Jolla, CA, USA*

CHRIS RIFE • *Joint Center for Structural Genomics (JCSG), Stanford Synchrotron Radiation Laboratory, Stanford University, Menlo Park, CA, USA*

EDWARD ROSS • *Biomedical Informatics Research Network, San Diego, CA, USA*

MARCO RUIZ • *Biomedical Informatics Research Network, San Diego, CA, USA*

BRIAN SANDERS • *The Center for Research in Biological Systems (CRBS), La Jolla, CA, USA*

MICHAEL R. SAWAYA • *Howard Hughes Medical Institute, UCLA-DOE Institute of Genomics and Proteomics, Los Angeles, CA, USA*

Chapter 1

Management of Information in Distributed Biomedical Collaboratories

David B. Keator

Summary

Organizing and annotating biomedical data in structured ways has gained much interest and focus in the last 30 years. Driven by decreases in digital storage costs and advances in genetics sequencing, imaging, electronic data collection, and microarray technologies, data is being collected at an alarming rate. The specialization of fields in biology and medicine demonstrates the need for somewhat different structures for storage and retrieval of data. For biologists, the need for structured information and integration across a number of domains drives development. For clinical researchers and hospitals, the need for a structured medical record accessible to, ideally, any medical practitioner who might require it during the course of research or patient treatment, patient confidentiality, and security are the driving developmental factors. Scientific data management systems generally consist of a few core services: a backend database system, a front-end graphical user interface, and an export/import mechanism or data interchange format to both get data into and out of the database and share data with collaborators. The chapter introduces some existing databases, distributed file systems, and interchange languages used within the biomedical research and clinical communities for scientific data management and exchange.

Key words: Neuroinformatics, Biomedical informatics, Distributed information management, Database, Data federation, Data warehouse, Information interchange, XML

1. Introduction

Organizing and annotating biomedical data in structured ways has gained much interest and focus in the last 30 years. Driven by decreases in digital storage costs and advances in genetics sequencing, imaging, electronic data collection, and microarray technologies, data is being collected at an alarming rate. The need to store and exchange such data in meaningful ways in support of data analysis, hypothesis testing, and future collaborative

Vadim Astakhov (ed.), *Biomedical Informatics,* Methods in Molecular Biology, vol. 569
DOI 10.1007/978-1-59745-524-4_1, © Humana Press, a part of Springer Science+Business Media, LLC 2009

use is pervasive. The push for structured storage of electronic information has been gaining much momentum in the medical field with the advent of electronic exchange formats like Health Level 7 (HL7), Clinical Data Interchange Standards Committee's (CDISC) operational data model, and the global push for the electronic medical record. Information powerhouses such as Google™ and Microsoft™ have become increasingly interested in applying their respective technologies to the electronic medical record. At the federal level, there have been increased numbers of solicitations for transdisciplinary and collaboratory projects. These projects require sophisticated informatics tools for data storage, retrieval, and exchange and are the main drivers behind increased use of distributed information resources in the scientific community. Because transdisciplinary projects rely on effective use of data from many domains, there is a genuine interest in the informatics community on how best to store and combine such data while maintaining a high level of data availability and quality.

2. Bioinformatics, Medical Informatics, and Biomedical Informatics

The term *informatics* describes a field that is focused on the structured storage, retrieval, and optimal use of information and data. Bioinformatics is defined in the National Institutes of Health's (NIH) technology glossary as "the study of the inherent structure of biological information and biological systems. It brings together biological data from genome research with the theory and tools of mathematics and computer science" (science.education.nih.gov/supplements/nih4/technology/other/glossary. htm). Generally the term *bioinformatics* is used to describe methods of information organization, analysis, and retrieval in general biology. The focus of bioinformatics has predominantly been on genetics, proteomics, and cellular data. Conversely, medical informatics describes the analysis and dissemination of medical data through the application of computers to various aspects of health care and medicine. The focus in medical informatics is on clinically specific information processing and exchange challenges in the realm of patient care. Biomedical informatics encompasses both medical informatics and bioinformatics. The field brings together bioinformatics, imaging informatics, clinical informatics, and public health informatics, covering the diverse domains of molecular and cellular research, tissue and organ research, patient research, and population research.

The specialization of fields in biology and medicine demonstrates the need for somewhat different structures for storage and retrieval of data. Although many of the needs overlap, many are different. For biologists, the need for structured information and integration across a number of domains drives development. There are many tools available on the Internet for genome-wide information and analysis systems and gene nomenclature. Some of the most popular are the National Center for Biotechnology Information (NCBI; http://www.ncbi.nlm.nih.gov/) databases, GeneCards®(www.genecards.org), the UCSC Proteome Browser (http://genome.ucsc.edu), the Ingenuity Pathways Analysis Systems (http://www.ingenuity.com/), HUGO gene nomenclature committee (www.gene.ucl.ac.uk/nomenclature), and the Gene Ontology consortium (GO; http://www.geneontology.org/). Each of these information and analysis systems provides a piece of the overall biological puzzle and provides ways of integrating information collected from them. For clinical researchers and hospitals, the need for a structured medical record accessible to, ideally, any medical practitioner who might require it during the course of research or patient treatment, patient confidentiality, and security are the driving developmental factors. Both de-identification and/or anonymization of human subject datasets, and when they are required, are debated topics at the federal level and directly impact information storage and exchange developers.

The Health Insurance Portability and Accountability Act of 1996 (HIPAA) sets a number of security, administrative, and privacy standards and safeguards for the transmission of protected health information (PHI). Security safeguards not only address confidentiality of datasets, where computer systems are located, how electronic media are disposed, and access control to information, but also the integrity of information that is transmitted or received. Administrative safeguards address the organizational integrity of an operation, including policies and procedures related to PHI, contingency plans, organization of security personnel, and adoption of administrative transaction code sets. Privacy standards and safeguards are presumably the most relevant HIPAA regulations for biomedical researchers. Specifically section CRF 164.501 of HIPAA sets forth conditions on how PHI can be exchanged or disclosed for research purposes. Research is defined in the standard as "a systematic investigation, including research development, testing, and evaluation, designed to develop or contribute to generalizable knowledge" *(1)* Researchers are allowed during the course of a study to obtain, disclose, and use PHI under the Privacy Rule with individual authorization or without individual authorization under limited circumstances. Some examples of the limited circumstances are authorization from an institutional review board (IRB), research on PHI of descendents, preparation for research such as designing a protocol that will later go before an IRB, and limited datasets with data use agreements.

For more information on the federal regulations describing PHI data usage visit the HIPAA website: www.hipaa.org.

3. Management of Data in Biomedical Organizations

Scientific data management systems generally consist of a few core services: a backend database system, a front-end graphical user interface, and an export/import mechanism or data interchange format to both get data into and out of the database and share data with collaborators.

3.1. The Database

A database is a software system for structured storage and retrieval of data. The database stores information in tables that are very similar to spreadsheets. Columns in the table store attributes associated with a particular record, stored in the rows. Storing information in a database provides many complex functions that would not be available from a simple Microsoft™. Excel spreadsheet such as queries aggregating data from multiple tables, complex mathematical expressions, and updating records in bulk, to name a few. Databases provide highly optimized query engines tuned for traversing extremely large amounts of data in a short time. The database itself is the engine behind the scenes. If you're not a database engineer or programmer you'll want an interface to interact with the database. Virtually any modern programming language could be used to develop the front-end interface to the database. Database interfaces come in two main flavors: thick and thin clients. Thick clients generally denote programs that are installed and run on the local computer. They are termed "thick" because they take up more space on your computer's hard disk, are usually platform dependent, and are not location independent. In order to use the thick client, one most often is sitting on the terminal where the software is installed. In contrast, a thin client refers to a web-based interface. These interfaces are deployed on a web server, a machine usually located elsewhere from the terminal one is using to interact with the web interface. Thin clients have a much smaller footprint than thick clients and generally only require a generic web browser. Most of the processing and interaction with the backend database occurs at the web server and not on the local machine. The local machine becomes a dumb terminal, responsible for displaying the pages sent from the web server and sending user information back to the server for processing. Recent improvements in web-enabled programming languages such as JavaScript (www.javascript.com), Asynchronous Javascript And XML (AJAX; www.ajax.org), and Extensible Application Markup Language

(XAML; www.xaml.net) have begun to move some of the processing burden off the web server, back to the user's terminal both to gain an improvement in speed and to relieve processing burdens on busy web servers. The most common user interfaces currently are web-based interfaces. Anyone who uses the World Wide Web has used a web interface interacting with a backend database. Wikipedia (www.wikipedia.org) is a good example of a thin client interface to a backend database. Wikipedia uses the MediaWiki software written in the PHP (www.php.net) programming language with a MySQL (www.mysql.com) or PostgreSQL (www.postgresql.com) backend database. All the pages and vast amounts of information available on Wikipedia are indexed and stored in the database. Another everyday example is Google's search engine. The thin client interface at www.google.com is a query mechanism into Google's vast archive of web pages.

3.2. Data Federations and Data Warehouses

Both how and where data gets stored in the data management system are equally important. Traditionally, large data warehouses had been used by industry and some scientific communities to store information for dissemination (MedInfo DDW, BrainMap, CCDB) *(2–4)*. A data warehouse is a central repository for storing information. The data is stored locally and provides a tightly controlled, curated, and relatively reliable environment for data access. Data warehouses work very well in industry where the data models rarely change and the information to store and retrieve is well defined. In the scientific community the data and usage models are quite different from those used in industry. In these domains the data models are constantly changing as technology progresses and different types of information need to be stored. In these domains the data warehouse becomes expensive to maintain because of continual growth and the vast amounts of related, but not identical, data that need to be stored. Unlike data warehouses, data federations are distributed. The databases are located in different places, running on independent servers, sharing no resources, and are connected by the network. The distributed databases are linked together by a common query mechanism and information is joined together, most often, by a query mediator.

Benefits of data federations include autonomous control over one's data, shared burden of hosting, storing, and maintaining the repository by participating sites, data replication, and data availability. In a data federation, each site maintains control over its data and decides what is shared to whom and when. Data federations usually provide support for data replication where one site replicates another site's data. Data hosting sites and users benefit from data replication. If the main site is down, users can gain access to those datasets from a replicant. When the main site comes back online, it can retrieve an up-to-date copy from a replicant. Difficulties with data federations include data availability, query performance, mediation between different database schemas, nomenclatures, and data coding standards. Because the data is not hosted in one database, it is much harder to guarantee stable data availability across the federation and quality of service usually suffers. Furthermore, care must be taken when designing a data replication scheme across the data federation to minimize data duplication while maintaining the best overall data availability and query performance. Query performance is dramatically impacted by geographically distributed data. A manuscript written in 2005 comparing federated search performance of an online federated clinical network receiving data from eight distributed information sources including the Internet sites Medline Plus and PubMed reported locally indexed databases returning results on an average of 0.061 sec whereas Internet sites were returning queries in the 3.76–4.55 sec depending on the source. The authors additionally found that the system time required to combine the results across information sources scaled linearly with the number of sources queried, ranging from 18.1 msec for a single source to 122.6 msec for nine sources *(5)*.

3.3. Information Interchange Languages

Data formats and mechanisms to get information into and out of the database are highly domain dependent. Different mechanisms are required for purely textual information compared to those needed for large amounts of binary information. In the medical informatics domain there has been an explosion of standards, terminologies, and infrastructures for sharing and distributing data: CDISC, CDASH, C/PDP, HISP, HL7, XML-HLOS, ANSI HITSP, IEEE-1073, ATA, HHS/ONC, CHI, AHRQ, caBIG™, BIRN, XCEDE, AIM, BRIDGES, SNOMED CT, LOINC, NPI, X12N. Each of these standards has strengths and weaknesses and each caters to a particular domain. Many data exchange formats that have emerged in the last 10 years have been specified using the eXtensible Markup Language (XML; www.w3.org/XML). XML is a text-based language for the structured storage of data. It has become very popular in data exchange specifications because it is customizable, structured, textual in nature, and can be self-describing depending on how one defines the markup tags of the schema. Since its creation in the mid 1990s, many additions have been made to the XML standard allowing for document validation, binary data encapsulation, and data range checking, increasing XML's expressional power and applicability to a wide number of domains. XML documents are structured into a hierarchical organization making them easy to parse and represent internally as tree data structures. Although there are many benefits of using XML to create data exchange languages, sometimes the overhead needed to parse and use XML documents is more than is really needed or the data is inherently binary and does not lend itself well to an XML representation. In the following sections a few data exchange languages will be discussed. Some are new and XML based while others have been in use for many years and are ASCII text and binary based. The reader should note that one size does not fit all and the importance of such exchange languages lies in the ability to store information in a well-documented and structured way, for use in information management, intra-application communication, and scientific data analysis. The languages discussed here are examples of packages in use and in development throughout various research consortia and clinical settings and provide the reader with a broad view of various flavors of data interchange languages.

3.3.1. XML-Based Clinical Experiment Data Exchange (XCEDE)

The XCEDE 2 XML schema (www.xcede.org), developed by members of the Biomedical Informatics Research Network (BIRN) test beds, provides an extensive metadata hierarchy for describing and documenting multisite human clinical and imaging studies. The schema is composed of hierarchically structured levels, storing information relevant to an aspect of an experiment (project, subject, protocol, etc.). Each hierarchy level allows for the storage of data provenance information providing a traceable record of processing and/or changes to the underlying data. The schema is extensible to support the needs of various data modalities and to express types of data not originally envisioned by the developers. An XCEDE 2 dataset is a data repository that can be represented as a collection of one or more XML files. The specification allows certain XML elements to link to other target elements, optionally specifying uniform resource locators (URLs) as hints to the location of documents containing those targets. For example, a given XCEDE 2 dataset may be stored as a single XML document, or a collection of files in a single directory on a file system, or may be distributed within a hierarchical directory structure (which may or may not reflect the structure of the data within), or may be stored within a database accessible by query through a web interface. The semantics of the dataset should be fully reflected in the XML representation, and should not be dependent on how the dataset is stored. The XCEDE schema was developed as an exchange language operating in a source–sink relationship with a data management system. As a source, the XCEDE formatted document serves as a structured, human-readable information exchange mechanism for deposition into a database. As a sink, the XCEDE document serves as an export mechanism to store data extracted from a database for analysis or exchange **(Fig. 1)** *(6).*

3.3.2. Clinical Data Interchange Standards Committee (CDISC)

CDISC is an open, multidisciplinary, nonprofit organization that has established worldwide industry standards to support the electronic acquisition, exchange, submission, and archiving of clinical trials data and metadata for medical and biopharmaceutical

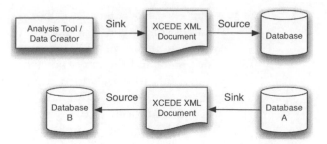

Fig. 1. XCEDE XML source–sink relationship with data creation tools and data management systems.

product development (www.cdisc.org). The CDISC consortium is made up of government agencies such as the Food and Drug Administration (FDA), European Medical Evaluation Agency (EMEA), National Institutes of Health/National Cancer Institute (NIH/NCI), and data management organizations such as the Society of Clinical Data Management (SCDM), the Association of Clinical Data Management (ACDM), and Data Management Biomedical (DMB). The CDISC standard is an XML-based standard and is separated into a number of models. The Operational Data Model (ODM) is a vendor-neutral, platform-independent format for data collected during the course of a clinical trial. The model provides structures for representation of study metadata, clinical data, and administrative data commonly collected during a clinical trial or archived after a clinical trial **(Fig. 2)**. The model complies with FDA 21CFR11 regulations. The Submission Data Standards model (SDS) contains guiding structures for the organization and tabulation of datasets acquired during a clinical trial and submitted to regulatory authorities such as the FDA. The Laboratory model (LAB) is a model for the acquisition and interchange of clinical trial laboratory data. LAB is a content model and the implementation can be through different data formats such as ASCII, SAS, XML, and HL7 (discussed below).

3.3.3. Annotation and Image Markup Project (AIM)

AIM is a project focused on semantic integration and annotation of medical imaging information *(7)*. The approach includes: *(1)* an ontology for describing the contents of medical images and an information model for semantic annotations; *(2)* an image annotation tool for collecting user annotations as instances of the ontology; *(3)* methods for transforming the annotations into common medical and Web accessible formats such as DICOM, HL7, CDA (XML), and OWL (www.w3.org/TR/2004/REC-owl-features-20040210/) *(8, 9)*. The ontology includes entities for representing associations in medical imaging derived anatomic structures and spatial regions with

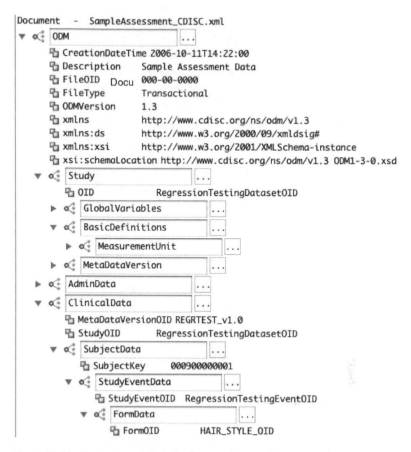

Fig. 2. CDISC ODM representation of clinical assessment data. The ODM XML instance document has sections describing the assessment questions and structure and sections documenting the values for each question collected from subjects.

observations made by radiologists. The anatomic structures in the ontology are taken from RadLex, a radiological lexicon designed to augment the American College of Radiology's terminology index with concepts related to anatomic and pathologic codes along with terms for devices, procedures, and imaging techniques, to name a few *(7, 10)*. The AIM project includes an information model describing the minimal information needed to annotate an image. The information model is represented in both OWL and XML schema to enable interoperability between hospital and web environments. To enable image annotations using the AIM ontology and information model, an image annotation tool is available which uses information from the ontology to determine the information collection requirements relevant to the region of the body and the qualities of various abnormalities, including size, shape, density, etc. For more information on AIM or to download the information model please visit www.gforce.nci.nih.gov.

3.3.4. Health Level 7 (HL7)

Hospitals and research centers typically use many different systems to track billing, patient records, and administrative records. In order for these various systems to communicate with each other effectively, a standard messaging format is needed. HL7 is an information exchange standard permitting structured exchange of clinical health care information between software systems. HL7's primary mission is to create flexible standards, guidelines, and methodologies to enable the exchange and interoperability of electronic health records (www.hl7.org). The "Level Seven" in the name refers to the seventh Open Systems Interconnection (OSI) layer protocol for the health environment and indicates that HL7's scope is the format and content of the data exchanged between applications instead of how the message is actually transmitted across the networks. The scope of HL7's activities includes modeling and methodology, vocabulary, clinical decision support, financial management, administration, regulated clinical research and information management, scheduling and logistics, clinical guidelines, community-based health, government projects, medication, security and accountability, and public health and emergency response. HL7 is unlike other information exchange languages in that it is not based on XML. HL7 V2.x messages have a variable-length, positional format and consist of lines ("segments") of ASCII text. Each line of text is a fixed sequence of fields or data elements separated by delimiters. In the HL7 standards document, each data item is well defined. HL7 V2.5 contains approx. 1,700 data items. Each data element is usually separated by vertical bar (or pipe "|"). Characters may have components (separated by "^" characters) and may repeat (e.g., for multiple patient IDs, phone numbers, etc.; **Fig. 3**). The HL7 standard has undergone many revisions since its beginnings in 1988. HL7 version 3 is currently under development and addresses a number of limitations with previous versions; namely, the need for an implicit information model, need for controlled vocabularies, need for explicit support of object technologies and security functions, need for limited single coding syntax, and tight coupling of events to profiles.

MSH|^~\&|PATH||GP123||200407161745||ORU^R01||101|P|2.5^AUS|34567||AL|NE|A

US||en<cr>PID|||KNEE123||Knees^Nobby^J^^Mr||19331215|M|||23 Shady

Lane^LIGHTNING RIDGE^NSW^2392|||||||219171803<cr>

OBR|1||PMS66666|956635.9|LFT^LIVER FUNCTION

TEST^N2270<cr>OBX|1|NM||751-7^S Albumin^LN||38|g/L|35-

45||||F<cr>OBX|2|NM||779-8^S Alkaline Phosphatase^LN||52|U/L|30-120||||F<cr>

Fig. 3. HL7 formatted message example of albumin and alkaline phosphatase levels in the liver.

Although HL7 is considered a standard, for historical reasons each vender implemented similar messages in slightly different ways. The result is a system that became less "standardized" than one would like. Changes in HL7 version 3 should alleviate some of the problems but because the new version will not be backwards compatible, hospitals and research centers will need to migrate all systems to the current version or run separate servers for each. Although HL7 is not an ideal electronic health record messaging standard, it has been adopted by many clinical applications and is improving workflows through out the healthcare industry by allowing applications to speak to each other.

3.3.5. Digital Imaging and Communications in Medicine (DICOM)

The DICOM digital imaging standard was born out of a collaboration between the ACR and the National Electrical Manufacturers Association (NEMA) in the 1970s after the introduction of numerous medical image sources to the marketplace and the use of computers to process those images (medical.nema.org). The original ACR-NEMA standard was published in 1988 and created a standardized terminology, an information structure, and unsanctioned file encodings. Standard methods for communicating digital image content across the networks were not completed until 1993 with the release of version 3.0 which accompanied the name change to DICOM. The DICOM standard addresses multiple layers of the OSI model, providing support for exchanging information over the network via TCP/IP, definition of operation Service Classes beyond simple data transfer, a mechanism for uniquely identifying Information Objects as they are acted upon across the network, and encoding the information for the applications. At the application layer, DICOM provides services and information objects for transmission and persistence of objects such as images and documents, query and retrieval support, workflow management, and quality and consistency of image appearance. The DICOM consortium currently supports 26 working groups focusing on various aspects of the standard ranging from data compression, structured reporting, and application support, to specific medical applications such as magnetic resonance imaging, clinical trials and education, surgery, and veterinary medicine.

The DICOM standard provides mechanisms for both modeling real-world objects and specifying services that use those objects. Each layer works together and ultimately presents the data across the network via message exchange or locally via a physical file (**Fig. 4**). The Information Object Definition (IOD) is an abstract object-oriented data model specifying information about real-world objects. The IOD is the formal representation of the data model, not a particular instantiation of such. The DICOM standard provides many IOD definitions. There are a number of relevant IODs for biomedical informatics applications spanning across all computed tomographic modalities, image registration,

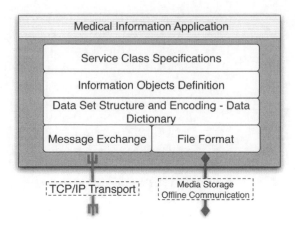

Fig. 4. General communication model for DICOM.

and interpretation, along with patient, visit, and study modules. The Service Class Specification associates one or more IODs with one or more services that act on them. The relationship of the IOD to the service class is somewhat analogous to data structure definitions in a C++ class and the functions in that class that operate on those data structures. Common service classes are storage, query/retrieve, workflow management, and printing. The DICOM data dictionary provides a centralized registry defining a collection of unique tag IDs defining elements and attributes. The data dictionary specification consists of a group and element number, a name, a value representation or data type, and a value multiplicity for encoding array structures. Theoretically, rigid adherence to a common data dictionary should make data interchangeable from application to application. What occurred in DICOM making instances of data less amiable to application sharing was the use of the private section. DICOM provides a private section outside the scope of the data dictionary. In the private section manufacturers could include tag elements with content not formally listed in the data dictionary. This functionality combined with the small percentage of DICOM "required" fields (to be considered DICOM compliant) created much difficulty for applications reading data originating from other vendors. Many times important information for reading and processing datasets were contained in the private tag sections making interchange very difficult. Since the early DICOM specification, the working groups have refactored the specification to require more data dictionary defined tag elements (to be DICOM compliant) and to rely less on private tags. The message exchange protocol and services provide a specification for exchanging messages and transmitting data over the network. Overall, DICOM has become an extremely pervasive standard, dominating the clinical medical imaging field as the de-facto standard for data transmission, storage, and exchange.

4. Informatics Components of a Federated Biomedical Data Management Environment

The development and deployment of a distributed infrastructure supporting collaborative biomedical imaging research integrates tools for data acquisition, data analysis workflows, and data retrieval/interpretation systems with the core infrastructure components and middleware. Common workflows in support of this endeavor consist of data import from scanning systems, storage of clinically relevant information in community or lab databases, preprocessing and analysis of imaging and clinical information, and storage of the derived data back into a data management system *(11)*.

4.1. Distributed Data Storage

Data storage in a distributed environment presents a number of challenges differing from share disk file systems, common in data warehouses. The storage space needs to be distributed across numerous computer systems, aggregating relatively small amounts of storage into a larger virtual space. There are many impediments to communicating with the distributed file system including access permissions, transport across firewalls, transfer speeds plagued by Internet traffic, and reliability to name a few. Security of the file system is also challenging in a distributed environment. How does a system guarantee that a client is actually authorized to access the data and how does a system prevent packets from being sniffed off the network? Each of these impediments can cause a number of related problems in the scientific data management system. The distributed file system, sometimes referred to as a networked file system or data grid, benefits from transparent data replication across sites and a level of fault tolerance. When a single node is offline, other nodes can host the replicated files. There are a number of open-source and commercial distributed file system packages currently available. The systems discussed here are examples of popular packages in use throughout various research consortia and do not represent an exhaustive list nor a detailed comparison of the pros and cons of any particular system.

4.1.1. Storage Resource Broker (SRB)

The Storage Resource Broker (SRB; http://www.sdsc.edu/srb/index.php/Main_Page) provides a uniform application programmer interface that can be used to connect to heterogeneous and distributed resources allowing users to seamlessly access and manage these datasets. Developed at San Diego Super Computing Center (SDSC), the software is a product of numerous research/development proposals focusing the development of SRB on applied research applications. The physical location of files is abstracted from the users and application programs, using a logical uniform resource identifier (URI) string to uniquely identify files across the distributed sites. The URI identifier is similar to WWW addresses commonly typed into a web browser. For example, the address www.google.com is a location-independent identifier for Google. When the web browser connects to the address www.google.com it is actually performing a lookup of which server it should contact with the uniform address www.google.com. In this way, Google can change the address, name, and location of their web servers without affecting how the public is connecting to the systems. In much the same way the SRB provides uniform access to files. Instead of asking for a particular file stored on a particular machine which would break if that machine were offline, one asks for a particular file and lets the SRB servers do the work of translating the URI string for that file into a transfer request to the particular system containing the file, or replicants of that system if it is offline, and returning it to the user. The SRB system provides functionality for investigators to establish a data hierarchy, and the appropriate access permissions, for multiple institutions and their researchers to contribute heterogeneous data to a shared system. Each participating institution has sufficient read and write access to the formal data hierarchy irrespective of their location and environment. Regardless of which site and what type of environment, the hierarchy looks and acts the same. While technologies such as iSCSI might perform faster than SRB, configuring mount points, access permissions, firewall rules for hundreds of users across 20 institutions would not be feasible for the types of projects that the SRB can handle today. In addition, the SRB supports file and directory metadata in the form of key-value pairs, allowing any relevant metadata to be attached and searched within the data grid *(24)*.

4.1.2. Andrew File System (AFS)

AFS is a distributed file system product, pioneered at Carnegie Mellon University and supported and developed as a product by Transarc Corporation (now IBM Pittsburgh Labs). It offers a client–server architecture for federated file sharing and replicated read-only content distribution, providing location independence, scalability, security, and transparent migration capabilities. AFS (http://www.openafs.org) is available for a broad range of heterogeneous systems including UNIX® (www.unix.org), Linux (www.linux.org), Mac OS X (www.apple.com/macosx), and Microsoft™ Windows (www.microsoft.com/windows). AFS works a bit differently than other distributed file systems. It creates a locally cached copy of files for increased speed and robustness when network access is temporarily interrupted. Read and write operations on files are performed on the local cached copy of the file. When a modified file is closed, the file is synchronized with the server, updating only the parts of the file that have changed. File consistency is maintained through system *callbacks*.

When a client machine caches a file locally, the server keeps track of this and sends the client a message if the original file has been changed by another user. This rather unique file locking mechanism prevents AFS from supporting large shared databases or record updating within files shared by many clients. AFS uses Kerberos (web.mit.edu/kerberos) authentication and maintains access control lists on directories for both users and groups. AFS has been designed to support large sets of users and has supported systems with over 50,000 users. AFS volume mounting works much like its Networked File System (NFS) counterpart. In fact, NFS was heavily influenced by AFS. Mounting of AFS file systems works in much the same way as NFS. System administrators create mount points on the local system which points to a named path in an AFS cell. After the mount point has been established, users read and write files in exactly the same way as with NFS or other network attached storage. AFS administrators control which server hosts the mounted volume and can change the physical location of the storage without affecting the client mount of the volume. AFS volumes can be used for read-only replication. As replicants are removed, the client is automatically redirected to the next replicant of the file. AFS administrators can also move replicants from one physical location to another without affecting the clients.

Spasojevic and Satyanarayanan did a performance evaluation of AFS over a 12-week period on 50 file servers and 300 clients from seven states. The analysis focused on both the quantitative and qualitative aspects of the file system. Qualitative questionnaires were distributed to participating sites eliciting user perceptions of performance and reliability, usage scenarios, access control lists, read-only replication, and data mobility. Quantitative measures were collected for both server and client machines spanning remote procedure calls responsible for data fetching, status, storing, removing, and creation. The average time to fetch data at the server was 116 msec and for the client, 158 msec. File storage at the server completed on average in 157 ms and for the client, 65 msec. The daily file transfer size distribution for server fetching was 156 MB and for client fetching, 5.3 MB. The daily file transfer size for storing at the server was 116 MB and at the client, 4.7 MB. For more detail on all the performance metrics collected and to get complete information on AFS as a distributed file system in practice please see (25) (http://www.usenix.org/publications/library/proceedings/sf94/full_papers/spasojevic.a).

4.1.3. CODA

Coda is a file system based on the AFS file system designed for large-scale distributed computing environments. Coda was developed to be substantially more resilient than AFS. Coda provides users with the benefit of a distributed file system while also allowing them to rely entirely on local resources when the client is disconnected from the network and to support distributed file system usage on portable computing platforms *(12)*. The Coda file system consists of client and server machines. A client connects to the "Coda" file system mounted under "/coda" and not to individual servers as with a Networked File System (NFS) mount. Unlike NFS clients that need to keep a list of servers and exported directories, Coda clients only need to know where to find the root directory "/coda" *(13)*. When new server shares are available, clients will discover these automatically in the Coda tree. Each client runs a process called Venus that handles remote file system requests and uses the local disk as a file cache. Clients cache files and directories in their entirety. Once the files are cached the client receives *callbacks* from the server in much the same way as AFS. Callbacks allow clients to maintain cache consistency while minimizing client–server interaction. Because files are cached locally, processes needing access to the files can operate on them as if they were on a regular Unix® file system. Processes at different sites see modifications to the file when it is opened or closed but not while edits are happening. Coda provides resiliency to server and network failures through the use of server replication. The unit of replication in Coda is a volume. A volume is a collection of files that are stored on one server and form a subtree of the shared space. The replication strategy is a read-one, write-all approach *(14)*. When the file is closed after being modified, it is transferred to all accessible members of an accessible volume storage group (AVSG). AVSG is the set of servers containing replicas of a volume. One advantage of this approach is that the burden of data propagation is on the client rather than the server. Another mechanism providing resiliency is disconnected operation, a mode of execution in which a caching site temporarily assumes the role of a replication site. Disconnected operation is a compromise between a completely server dependent distributed file system and one that is completely autonomous. Disconnected operation is particularly useful for supporting portable workstations. Coda clients have access to all files that were cached before the network was disconnected. Once the system is reconnected to the network, modified files and directories are propagated to the AVSG. Under normal circumstances the Coda least recently used (LRU) caching mechanism is sufficient. If a client is expected to be disconnected for extended periods of time, an LRU cache may not be sufficient to guarantee the needed files are available. Coda therefore allows the user to specify a priority list of files and directories that should be retained in cache. Objects with the highest priorities become sticky and are retained at all times. Additionally, Coda allows users to bracket a sequence of high-level actions and will record the file references generated during those actions and maintain those files in the cache for offline use.

One place where Coda might be extremely useful is in World Wide Web (WWW) server replication *(13)*. Many Internet service providers and/or large web sites have too much traffic to be handled by a single server. The servers are therefore replicated and traffic is load balanced between each server. Many files and documents potentially need to be shared between all the servers. Site administrators have attempted to use NFS to solve the issue but performance has proven problematic. Typically files are copied to the local disk of individual servers. Coda's file caching functionality might work nicely in this environment. When the files are cached they are effectively local files and performance should not be dramatically affected. For more information on Coda and to download the software and get up-to-date information, please visit www.coda.cs.cmu.edu.

4.1.4. Ceph

Ceph is a distributed file system designed for high-performance computing applications where storage sizes are in the petabyte range. It provides excellent performance, reliability, and scalability. Ceph decouples data and metadata management into separate clusters allowing distribution of the complexity involved with data access, update, replication, failure detection, and recovery. Metadata servers (MDS) interact with clients to support operations such as opening and renaming files whereas for read and write operations the client interacts directly with intelligent object storage devices (OSDs) *(15)*. Most distributed file systems use a static subtree partitioning scheme where the administrators set up volume partitions ahead of time based on predicted loads. Ceph utilizes dynamic subtree partitioning to effectively repartition how metadata is represented across the MDS cluster. As loads increase for particular types of metadata operations, the MDS adaptively redistributes metadata across the MDS nodes to keep workloads evenly distributed. Directories that are heavily read are replicated across multiple nodes to distribute the load. Directories that are large or are experiencing heavy writes have their contents hashed by file names across the cluster, balancing the distributions. OSDs are more than just spinning disks, they combine a CPU, network interface, and local caching with the underlying spinning disks in various RAID configurations. OSDs are responsible for data migration, replication, failure detection, and failure recovery while providing a single logical storage device to clients. OSDs do not directly interact with the MDS, all interaction occurs between the client and each system. If one or more clients open a file for read access, an MDS grants them the capability to read and cache the file by returning the inode number, layout, and file size to the client. The client can then locate the file data and read it directly from the OSD cluster. In Ceph there is no need for file allocation metadata; the inode number and the stripe number are all a client needs to read the file from the OSD cluster. The CRUSH data distribution function takes care of assigning objects to storage devices in the OSD cluster. The MDS cluster manages client interaction with the file system. Both read and updates are synchronously applied by the MDS to ensure serialization, consistency, security, and safety. No callback system is used as in AFS and Coda. Weil et al. collected performance metrics on a 430-node partition of the alc Linux cluster at Lawrence Livermore National Laboratory and found that as the MDS cluster grows to 128 nodes, the efficiency drops no more than 50% below perfect linear scaling for most workloads showing vastly improved performance over existing systems *(15)*. For more information on Ceph and full discussion of the performance metrics please visit ceph.sourceforge.net/index.html.

4.2. Distributed Biomedical Data Management Systems

Data management systems specifically designed to operate in distributed environments are few and far between in the biomedical research community. Most collaboratories have adopted the data warehousing approach with a single, centralized database and a web-accessible interface. There are a number of tools in development for information integration from databases that are not specifically designed for data federation. Grid-enabled, large-scale collaboratories such as BIRN (www.nbirn.net), caBIG (cabig. nci.nih.gov), and BRIDGES (www.brc.dcs.gla.ac.uk/projects/ bridges) *(11, 16)* approach the problem of data federation in different ways. Some are focusing on developing query mediators and schema mapping functionality, others are using commercial products like IBM's Information Integrator (www-306.ibm.com/ software/data/integration/) to combine data, while still others are focusing on using standard rules and common languages to build interoperable systems. Another interesting approach found in the literature is the use of content management systems (CMS) to streamline developing shareable information resources, which with some extension could be federated *(17)*. The data management systems and content frameworks discussed here were selected to illustrate approaches to data management used in currently active collaboratories focused on biomedical research.

4.2.1. Human Imaging Database (HID)

The HID is an open-source, extensible database schema implemented in Oracle (www.oracle.com) 10g, 9i, and PostgreSQL (www.postgresql.org) 7.x, 8.x, and associated three-tier J2EE application environment for the storage and retrieval of biomedical data designed to operate in a federated database environment *(18)*. The HID was designed as an open-source data management system for the Biomedical Informatics Research Network (www.nbirn.net) *(11)*. The HID at a particular site contains relevant information concerning the research subjects used in an experiment, subject assessments, the experimental data collected, the experimental protocols used, and any annotations or statistics normally included with an experiment. The database is composed of an extensible schema and structured core. The core database contains a hierarchical description of an experiment and how experimental protocols are related. Each descriptor in the database consists of a "base tuple" that defines the minimum informational requirements of that descriptor. For example, the core database schema has no concept of genetic information. If a lab decided to start collecting genotype data on subjects, a "base tuple" descriptor would be created in the database, called genotype for instance, and the specifics of the data (i.e., chromosome location, RS#, genotype) actually collected would be stored in the extended tuples tables of the schema, indexed by the "base tuple." The database can therefore be extended for various experiments without changing the underlying schema and those extensions can be reused and/or modified for other experiments. In addition, the database contains an extensible framework for the definition and storage of clinical assessment and demographic data. This dedicated section of the schema handles the storage of assessment data from a wide variety of assessments and their modifications (i.e., assessment version control). Information is stored that allows for the annotation of the status or quality of assessment data. All missing data can be coded to differentiate between data that was not entered for unknown reasons and those that the subject declined to answer or other possibilities that can affect the interpretation of resulting analyses. Some of the data quality measures put in place for clinical data include double-entry, data type and range validation, and manual validation and logging for those data which fail the double-entry, prior to being exposed to a query service. The data management system comes with a graphical user interface for rapid assessment data entry form creation. The Clinical Assessment Layout Manager (CALM) is a tool enabling drag and drop data entry form creation without programming. The forms designed in CALM are deployed on the web server and metadata is inserted in the HID database directly from CALM providing seamless data entry and query support for newly designed forms. The HID environment also contains an intuitive web-based user interface for the entry and management of subject's data and a data integration engine, allowing multiple sites running the HID to create a federated database such that these sites can be queried as a single database resource from the web-based user interface. All that is required to facilitate data federation is to "tell" each HID instance about all other HID instances it can access. The user then chooses to perform a pseudo-mediated query or a single site query through the web interface. There are currently 12 federated HID databases deployed in BIRN, 10 Oracle and 1 PostgreSQL versions, storing clinical information on more than 419 subject imaging visits and 3,174 subject assessments. For more information, please visit www.nitrc.org/projects/hid.

*4.2.2. The Extensible
Neuroimaging Archive
Toolkit (XNAT)*

XNAT is an open-source, Java-based, extensible data management system designed to serve as a hub for capturing data from multiple sources, archiving, and distributing it to collaborators *(19)*. XNAT's workflow model is most closely related to a central site data warehouse. Raw data and newly processed data are imported directly to the data management system from the data acquisition system. The data is then quarantined and reviewed for data quality. Once the data has passed QC it is released for local use. The data is then processed and analyzed using available computational resources. Raw and processed data are then made available for collaborative use, as permission to projects and data are granted to a widening audience. Finally, prepared datasets are anonymized and made available for public use at centralized data centers. XNAT has an extensible data model, modular design, and customizable user interface creating a flexible platform that can be used for a wide range of research domains. XNAT is unique from other data management systems presented here in that it relies heavily on XML and XML Schema for its data representation, security system, and generation of user interface components. One of the most dramatic differences between XNAT and other data management systems is the lack of a rigid database schema. The system provides a number of built-in data types and catalogs to help laboratories model their own data in XML. Once a data model is created in XML, XNAT uses the data model to build database schema tables, setup relationships, triggers, and all the behind-the-scenes functionality contained in data management systems. XNAT's user interface design functionality allows the user to modify the user interface components and select which functionality to include in the database instance. The system comes with a programmer's interface providing developers with the tools to implement more sophisticated functionality. Although XNAT was originally designed as a single-site data archiving toolbox, it has been deployed within the BIRN collaboratory as a shared resource for many years and work is ongoing to support a more sophisticated web services layer based on XCEDE, effectively exposing the data archive to a larger federated data environment. XNAT's XML-based design was deliberately chosen to simplify transitioning from a single-site data archive to a multisite distributed environment. For more information please visit www.xnat.org.

*4.2.3. Cancer Central
Clinical Database (C3D)*

C3D is a clinical trials data management system created by the Cancer Biomedical Informatics Grid (caBIG™; cabig.nci.nih.gov). caBIG™ is an information network enabling researchers to share data, develop standards and a common language for data sharing, and build or adapt tools for collecting, analyzing, integrating, and disseminating information associated with cancer research. The caBIG™ network is by definition federated in that each site maintains control over their data and resources and by using a common underlying framework becomes interoperable.

C3D is the web-based clinical data management solution designed on top of caBIG's core infrastructure. It was originally developed for the National Cancer Institute's intramural research program and is now made available to other research organizations in a hosted Application Service Provider model. C3D supports data standardization, reuse, sharing, and interoperability through electronic Case Report Forms (eCRFs) based on Common Data Elements (CDEs) maintained in NCICB's Cancer Data Standards Repository (caDSR) and controlled by terminology from the NCI Enterprise Vocabulary Services (EVS) *(20)*. C3D collects clinical trial data using standard case report forms (CRFs) based on CDEs. C3D utilizes security procedures to protect patient confidentiality and maintain an audit trail as required by FDA regulations. C3D currently supports electronic submission of clinical trials data to the National Cancer Institute's (NCI) Clinical Data System (CDS) and the Clinical Trials Monitoring Service (CTMS/Theradex). C3D consists of three web-based components: Oracle Clinical, for protocol building; Remote Data Capture, for data entry and management; and Integrated Review/Java Review, for real-time access to clinical data within and across clinical studies to author-ized users. C3D is currently managing over 160+ production studies containing 6100 patients and more than 2.8 million lab results. Although C3D is not an open-source product and is built around Oracle, a proprietary database engine, it has been used widely in the cancer network to effectively manage data from multiple sites in a distributed environment. For more information please visit cabig.nci.nih.gov/tools/c3d.

4.2.4. Content Management Based Systems (CMS)

Content management at a high level is defined as the process of collecting, managing, and publishing content, making it available to users through any outlet *(21)*. Some common forms of content available on the World Wide Web today are Hyper Text Markup Language (HTML) pages, XML files, images, audio, videos, documents, email, technical documents, e-commerce transactions, and dynamic content from relational databases *(22)*. Over the last 15 years the Internet has seen explosive growth in the number of web sites for content management. Similarly, there has been an explosive growth in content management solutions such as MediaWiki (www.mediawiki.org), Zope (www.zope.org), KnowledgeTree (www.knowledgetree.com), Plone (www.plone.org), Struts2 (struts.apache.org), and many others. After close inspection of these content management solutions one finds that many of the same components required for a web-based scientific data management system are part of the CMS systems. For example, CMS systems have a web interface of some flavor, a backend database platform for content management, an API for customi-zation, and a security layer (some more sophisticated than others) for user management. Because the CMS tools handle the nuts and bolts of web application and document management, groups can put more effort into developing the specific components needed to facilitate scientific data sharing and less on the overall development of the CMS system *(17)*. In Mooney's 2008 article on using CMS systems to support biomedical science, the author suggests that although CMS systems have not often been used for biomedical data, the architecture has potential as a solution for managing shared, accessible data. Indeed all of the large-scale collaboratories mentioned in this chapter, BIRN, caBIG™, and BRIDGES, maintain a content management system of some kind to support collaboration. In addition, it is interesting that the framework behind the web interface of the BIRN HID, discussed previously in this chapter, is based on Struts. Why couldn't these same systems, with some modification, be used for scientific data management without the need to reinvent the wheel, designing custom data management solutions? In an article published in Nature in 2007 by Jim Giles there is an example of this very thing, termed the Wiki for Professionals (www.wikiprofessional.info) *(23)*. The site integrates information from the Swiss-Prot (www.psc. edu/general/software/packages/swiss/swiss.html) protein database, with genetic information from the Gene Ontology (www. geneontology.org/) database, and disease information from the US National Library of Medicine into a single resource. The site further allows registered researchers to modify, edit, annotate, and add content dynamically to the site. Although this example is a bit different in scope because it is not focused on sharing human data which comes with all sorts of regulatory caveats, it is an interesting example of how a collaboratory could use CMS systems that were not specifically designed to operate in the realm of science to integrate information from multiple data sources and possibly, with some additional work, support a data federation. In an age where federal dollars earmarked for scientific research are shrinking, using CMS systems as the basis for data management and collaborative web-enabled solutions could save vital time and money in developing custom solutions on a per project basis along with facilitating collaboration and data sharing.

Acknowledgment

This work was supported in part by [5 U24 RR021992] to the Functional Imaging Research in Schizophrenia Testbed (FIRST) Biomedical Informatics Research Network (BIRN, http://www. nbirn.net). The author thanks the University of California, Irvine Center for BioMedical Informatics for discussions on clinical information systems and the informatics needs to support clinical translational research.

References

1. Title 45, Volume 1, Parts 1 to 199. 1997. (Accessed at http://www.hhs.gov/foia/45cfr5. html.)

2. Wu C. Development of a medical informatics data warehouse. AMIA Annu Symp Proc 2006:1148.

3. Laird AR, Lancaster JL, Fox PT. BrainMap: The social evolution of a human brain mapping database. Neuroinformatics 2005;3(1):65–78.

4. Martone ME, Tran J, Wong WW, et al. The cell centered database project: An update on building community resources for managing and sharing 3D imaging data. J Struct Biol 2008;161(3):220–31.

5. Coiera E, Walther M, Nguyen K, Lovell NH. Architecture for knowledge-based and federated search of online clinical evidence. J Med Internet Res 2005;7(5):e52.

6. Keator DB, Gadde S, Grethe JS, Taylor DV, Potkin SG. A general XML schema and SPM toolbox for storage of neuro-imaging results and anatomical labels. Neuroinformatics 2006;4(2):199–212.

7. Rubin DL, Mongkolwat P, Kleper V, Supekar K, Channin D. Medical Imaging on the Semantic Web: Annotation and Imaging Markup. 2008. AAAI Spring Symposium Series, Semantic Scientific Knowledge Integration 2008.

8. Mildenberger P, Eichelberg M, Martin E. Introduction to the DICOM standard. Eur Radiol 2002;12(4):920–7.

9. Quinn J. An HL7 overview. J AHIMA 1999; 70(7):32–4.

10. Langlotz CP. RadLex: A new method for indexing online educational materials. Radiographics 2006;26(6):1595–7.

11. Keator DB, Grethe JS, Marcus D, Ozyurt B, Gadde S, Murphy S, Pieper S, Greve D, Notestine R, Bockholt HJ, Papadopoulos P; Function BIRN, Morphometry BIRN, BIRN-Coordinating Center. A National Human Neuroimaging Collaboratory Enabled by the Biomedical Informatics Research Network (BIRN). IEEE Trans Inform Technol Biomed Special Bio-Grid edition 2008;12: 162–72.

12. Satyanarayanan M. Scalable, secure, and highly available distributed file access. IEEE Comput 1990;23(5):9–25.

13. Braam P. The Coda distributed file system. Linux J 1998;50.

14. Satyanarayanan M. Coda: A highly available file system for a distributed workstation environment. In: IEEE Workshop on Workstation Operating Systems; Pacific Grove, CA; 1989.

15. Weil SA, Brandt SA, Miller E, Long D, Maltzahn C. Ceph: A scalable, high-performance distributed file system. In: Conference on Operating Systems Design and Implementation; 2006.

16. Sinnott R, Gilbert D, Berry D, Hunt E, Atkinson M. Bridges: security focused integration of distributed biomedical data. In: UK e-Science All Hands Meeting; UK; 2003.

17. Mooney SD, Baenziger PH. Extensible open source content management systems and frameworks: A solution for many needs of a bioinformatics group. Brief Bioinform 2008;9(1):69–74.

18. Ozyurt B, Wei D, Keator D, Potkin S, Brown G, Grethe J. Web-accessible clinical data management within an extensible neuroimaging database. In: Society for Neuroscience; Washington, DC; 2005.

19. Marcus DS, Olsen TR, Ramaratnam M, Buckner RL. The Extensible Neuroimaging Archive Toolkit: An informatics platform for managing, exploring, and sharing neuroimaging data. Neuroinformatics 2007;5(1): 11–34.

20. Diercksen K. Cancer Central Clinical Database (C3D) Overview; 2008.

21. Boiko B. Understanding content management. Am Soc Inf Sci 2001;28(October-November): 8–13.

22. Frost S. Web Content Management Report. London; 2001.

23. Giles J. Key biology databases go wiki. Nature 2007;445(7129):691.

24. Rajasekar A, Wan M, Moore R, Schroeder W, Kremenek G, Jagatheesan A, Cowart C, Zhu B, Chen S, Olschanowsky R. Storage Resource Broker-Managing Distributed Data in a Grid. Computer Society of India Journal, Special Issue on San. 2003;33(4):42–54.

25. Spasojevic M, Satyanarayanan M, An Empirical study of a Wide-Area Distributed File system. ACM Transactions on Computer systems (TOCS). 1996;14(2):200–222.

Chapter 2

Enabling Public Data Sharing: Encouraging Scientific Discovery and Education

Christine Fennema-Notestine

Summary

To promote scientific discovery and education, the federated Biomedical Informatics Research Network (BIRN) Data Repository (BDR) supports data storage, sharing, querying, and downloading for the biomedical community, enabling the integration of multiple data resources from a single entry point. The BDR encourages data sharing both for investigators requesting assistance with databasing and informatics infrastructure, and for those wishing to extend the reach of existing data resources to be registered with the BDR. Both approaches rely heavily on data integration and knowledge management techniques, ensuring capabilities for intelligent exploration of diverse data resources that make up the BDR's shared environment. Although the development of the BDR has been driven by BIRN testbeds in the fields of neuroscience and neuroimaging, the infrastructure is flexible and extendable to serve a broad array of disciplines, fueling interdisciplinary studies. For a welcoming environment, the BDR provides simple, straightforward policies and procedures for contributing data and for using available data. Contributing investigators provide information through the BDR Gridsphere-based Portal environment to allow meaningful sharing of their data, and, when relevant, supply documentation for human subjects' protection. The BDR aims to provide a common resource to increase the availability of, and access to, rare data; complex, sizable data; and existing data sharing structures in support of research and education in the scientific and clinical communities.

Key words: Data sharing, Collaborative environment, Repository, Scientific discovery, Data integration, BIRN

1. Introduction

Bioinformatics has catalyzed the remarkable growth of collaborative environments for the scientific research community to include the sharing and exchange of raw data, analysis tools, and derived or analyzed data *(1–7)*. Of particular relevance to a wide variety of individuals, public data sharing has the power to transform

Vadim Astakhov (ed.), *Biomedical Informatics,* Methods in Molecular Biology, vol. 569
DOI 10.1007/978-1-59745-524-4_2, © Humana Press, a part of Springer Science+Business Media, LLC 2009

modern science by facilitating scientific discovery and education. The NIH, in fact, expects a data sharing plan for most scientific applications, and strongly encourages "making data as widely and freely available as possible, while safeguarding the privacy of participants, and protecting confidential and proprietary data" *(8)*. The creation of a broad-based, federated data repository that enables intelligent search across multiple resources, which are comprised of a variety of data collected from different species, will drive development of new hypotheses and provide an abundance of collections for training research and clinical scientists.

An infrastructure that allows for public exchange of data enables new research exploration *(9)*, as investigators from different disciplines view riches in data acquired for a different purpose *(10)*. A public data repository encourages scientific inquiry by enabling investigators with diverse backgrounds to examine data related to their own area of specialty, to pool data across multiple sites *(11, 12)*, and to study rare data made available to a wider audience *(13)*. Investigators who may not have specialized resources, such as computational power for intensive analyses, novel imaging techniques, or access to clinical populations, are able to address broader-reaching questions relevant to their scientific hypotheses *(14)*. The power of such a repository is demonstrated by increased sensitivity in investigations where data may be combined, and by providing a common resource for test and training data to enhance methods development. Furthermore, the accessibility of resources, and advancements derived from these resources, will promote education and help train multi- and interdisciplinary investigators, students, and research teams, providing a resource complementary to clinical inquiry, data collection, data management, and ethics training *(15)*. Data sharing promotes many goals of the NIH research endeavor.

In this context, the NCRR-sponsored Biomedical Informatics Research Network (BIRN) *(7, 16)* has designed a federated informatics infrastructure to support storage, sharing, querying, and downloading of data for the biomedical community, integrating multiple data resources from a single front end. This BIRN Data Repository (BDR) aims to provide a common resource to increase the availability of, and access to, rare data sources, complicated imaging data resources, and existing well-defined data sharing efforts. The BDR encourages data sharing both for investigators wanting to contribute to the established databasing and informatics infrastructure, as well as those wishing to extend the reach of existing data resources through registration with the BDR. Both methods of sharing rely heavily on data integration and knowledge management techniques, ensuring the interoperability of the diverse data sources that make up the BDR's shared environment, and allow for data discovery.

The development of the BDR has been driven by the BIRN testbeds emphasizing clinical neuroscience research in humans and animal models. The complexity of the structure and function of the brain, in concert with behavioral, clinical, psychiatric, and genetic variables, has provided the ideal environment for the development of integrated data resources. Disparate shared human and animal model data resources incorporate image volumes, image metadata, data provenance, demographic, clinical assessments, behavioral data, experimental design, protocol details, microarray data, and derived values resulting from data analyses. Notably, the federated BDR infrastructure has been developed to be flexible and extendable to serve a broad array of disciplines, fueling interdisciplinary studies. Such infrastructure is critical to enable meaningful data sharing and requires continued maintenance and improvements as technologies and community needs change over time. For a welcoming environment, the BDR provides simple, straightforward policies and procedures for contributing data and for using available data. Contributing investigators provide minimal information through the BDR Portal environment, powered by Gridsphere *(17)*, to allow meaningful sharing of their data, providing documentation for human subjects' protection when appropriate.

2. Methods

Through data sharing working groups across the multisite BIRN initiative and review of institutional and existing repository policies, we sought to balance ease of sharing with the need for sufficient information to make shared data as useful to the community as possible. As the party responsible for the integrity of the data to be shared, the Contributor initiates the sharing process through the BDR Portal and provides a high-level description of the study to illustrate the context in which the data were originally acquired *(2.1)*. The shared data remain linked to the initial data submission and allow Users to browse the data source information along with the data (*2.4*). A delineation of the data types to be shared is requested to determine the minimum associated metadata to make the data meaningful to the public; the databasing or data integration requirements; and whether the data are potentially subject to human subjects' protection policies *(2.1)*. Finally, the BDR staff works with the Contributor to ensure that data are provided in readable formats and, when necessary, that data are sufficiently de-identified (*2.2, 2.3*).

For all data to be shared, BDR policy requires the Contributor to provide a minimum set of associated metadata to sufficiently describe his or her data, to ensure that shared data is meaningful and useful. These minimal information requirements are defined by the type of data and species, and are verified during the *Curation and Annotation* processes (*2.3*). For example, to share anatomical MRI data, the acquisition protocol and any preprocessing steps applied to these data (i.e., data provenance) are provided. Depending on the image file format, much information may be machine readable from the image header (e.g., TR of acquisition); additional information is provided by the Contributor through the Portal. To accommodate a wide variety of research areas, the BDR accepts commonly readable data formats, provided that sufficient descriptive information is furnished.

Data integration through the federated BDR environment relies heavily on data annotation *(18)*. Contributors work with the BDR to annotate and map shared data and database tables to a common lexical and ontological framework (BIRNLex; *2.3*) to allow for intelligent search and integration with other data sources. Finally, the Contributor agrees to make these data publicly available in a reasonable time frame, as determined by the nature of the data and collaborative investigations (*2.4*).

2.1. Data Management Interface

A rich curatorial environment, built on the BIRN Portal foundation powered by Gridsphere *(17)*, manages data submissions and the subsequent sharing of data. The BDR Portal provides a standardized interface for Contributors to provide details about their data to be shared, submit IRB-related documentation and approval when appropriate; access information related to data de-identification and upload or database registration; curation and annotation of data; and release of the final product. To maintain a BDR Portal account, the Contributor provides standard contact information and acceptance of the standard BDR Data Use Agreement policy (*2.4*). The Portal account provides a secure, web-based interface for a Contributor to work with the BDR staff toward sharing their data.

To create a new submission, the Contributor provides a title, brief abstract, and other information describing the original study; keywords selected from existing ontological resources; and, as relevant, co-investigators, affiliated institutions, sponsors, and funding sources. Following this brief initial step, the submission is assigned an accession number that will follow any of these data throughout the submission and subsequent sharing process. As these data are shared, Users presenting and publishing studies agree to acknowledge this original data source through this accession number (*2.4*).

The submission process may be resumed by either the original Contributor or a confirmed designee. That is, the Contributor may allow another individual to complete the process through his own Portal account; this shared project will be available for access to both individuals. The Contributor then provides details related to the data to be shared, including data types (e.g., MRI imaging, functional MRI, demographic information, cognitive measures) and metadata information to make shared data sufficiently meaningful (e.g., number of cases, species, sex, age range, and diagnostic groups). A declaration whether data to be shared is human or nonhuman follows; for human data sharing, steps are invoked related to human subjects' protection (*2.2*). Finally, information on how the data is currently stored is provided to initiate discussion on databasing strategies. Contributors may upload or import their data into existing BDR resource databases or work with the BDR staff to register and integrate their existing database with the federated system. Data can be imported into one of several database options common to the BDR, such as the Cell Centered Database (CCDB) *(14)*, Extensible Neuroimaging Archive Toolkit (XNAT) *(19)*, Human Imaging Database (HID) *(20, 21)*, or Contributors may wish to proceed with steps to register and annotate an existing database with the BDR. Finally, based on the nature of the project, a timeline for the public sharing of data is determined, based on estimated time for curation and data annotation. Additional factors may influence the time to sharing, if collaborative projects are underway, such that time to publication may lengthen the duration prior to sharing.

During this process, the BDR Submissions staff manages submitted information through the Portal, updates project summaries available through Contributors' accounts, and notifies Contributors of steps remaining to complete the submission. Notifications occur as alerts containing messages in the Portal account as well as automated, simultaneous e-mail contact directly to the Contributor with links back to the project for simplicity.

2.2. Human Subjects' Data

For the contribution of human subjects' data, the BDR requires data de-identification and verification of the Contributor's IRB approval and/or waiver for sharing such data with the broader community. The BDR staff are available to assist the Contributor in this process and provide suggested language to submit to her local IRB. The standard language assures the local institution that the data will be shared without personal identifiers according to HIPAA Privacy Rule regulations, and requests a waiver of consent to share these de-identified data. That is, through collaboration with the BDR, local data, including image header files, must be de-identified prior to upload into the BDR. This includes the removal of identifying information such as name, date of birth, and medical record number. Confirmation of de-identification is completed in the secure Portal environment prior to sharing with the public (*2.3*). The IRB language also states that nonidentifiable BIRN IDs are assigned to each subject prior to submission to the BDR; only the Contributor's local study team maintains the link table between internal subject identifier and the unique subject ID number assigned by the BDR. Additional suggested language is provided to the Contributor to share details about the study cohort and data types with the IRB.

The steps required by each local institution and the result of the request, including additional requirements for sharing, may depend on the nature of the study and how the data were collected. Electronic copies of the IRB documentation for the sharing of human data also will be uploaded to the BDR as verification of approval.

2.3. Curation and Annotation

Contributors are expected to adhere to a "minimal information" framework. That is, Contributors supply a minimum set of associated metadata to sufficiently describe submitted data, to ensure that shared data are meaningful and useful to the scientific community. The minimum set is determined by the nature of the data being shared, and, to some extent, is defined on a case-by-case basis. BDR staff will review the uploaded data and minimal information provided and will work with the Contributor to finalize the dataset to be shared. This period of data curation confirms de-identification of any human data, confirms the integrity of uploaded files, assesses the database results, and annotates the data for intelligent search across the BDR.

To facilitate interoperability, all data within the BDR is mapped to the shared terminology framework (BIRNLex/Ontology) *(22–25)* for data integration. This concept mapping allows the implementation of intelligent searches across all data available through the BDR; these data are stored in a variety of databases linked together to allow cross querying and data integration. The BDR staff works with the contributing individuals to annotate shared data utilizing a Concept Mapper tool. This tool provides a list of unique data values for each data entity; all instances of the unique values are automatically mapped to the same ID. If a precise match in terminology is not available (i.e., the term has not been previously entered and defined in BIRNLex), new terms are incorporated through a standardized template created collaboratively with the Contributor's team and the BDR staff. Throughout this process, the BDR staff assists the users and provides the tools required to annotate their data.

Once curation and annotation are complete, the Contributor receives BDR confirmation via e-mail and her Portal account. If the curation and upload are successful, a summary of the data submission is provided for review. The BDR also provides a link to a survey for User feedback and DVD copies of the public version of data for archival purposes. The timeline for release of these data to the public is based on the Contributor's formal agreement with the BDR.

2.4. Data Use and Discovery

The Data Use Agreement for the BDR is meant to encourage the public use of data in the repository, requiring little from potential Users. To download data from any resource in the repository, the User must agree to make no attempt to determine the identity of any individual; to employ data at his own risk as there is no guarantee provided by the BDR as to the quality of the data; and to acknowledge the use of data from the BDR in any public presentation (including but not limited to papers, books, book chapters, conference posters, and talks). The agreement provides specific wording for methods and acknowledgment sections, which includes a link to the BDR Project Accession Number identifying the original source of the data and reference to the BDR and associated funding resources. In addition, the User agrees to provide the BDR with a bibliographic citation of the final published presentation or article for inclusion in the BIRN literature archive. These same guidelines apply to any redistribution of these data or data derived from these data.

Acknowledgment

The Biomedical Informatics Research Network (BIRN) Data Repository (http://www.nbirn.net/bdr) is supported by grants to the BIRN Coordinating Center (U24-RR019701), Function BIRN (U24-RR021992), Morphometry BIRN (U24-RR021382), and Mouse BIRN (U24-RR021760) Testbeds funded by the National Center for Research Resources at the National Institutes of Health, USA.

References

1. Barrett T, Troup DB, Wilhite SE, et al. NCBI GEO: mining tens of millions of expression profiles – database and tools update. Nucleic Acids Res 2007;35:D760–5.

2. Berman H, Henrick K, Nakamura H, Markley JL. The worldwide Protein Data Bank (wwPDB): ensuring a single, uniform archive of PDB data. Nucleic Acids Res 2007;35:D301–3.

3. Bogue MA, Grubb SC, Maddatu TP, Bult CJ. Mouse Phenome Database (MPD). Nucleic Acids Res 2007;35:D643–9.

4. Breitkreutz BJ, Stark C, Reguly T, et al. The BioGRID Interaction Database: 2008 update. Nucleic Acids Res 2008;36:D637–40.

5. Neuroimaging Informatics Tools and Resources Clearinghouse (NITRC). 2006. (Accessed at http://www.nitrc.org/.).

6. Internet Analysis Tools Registry (IATR). (Accessed at http://www.cma.mgh.harvard.edu/iatr/.)

7. Grethe JS, Baru C, Gupta A, et al. Biomedical informatics research network: building a national collaboratory to hasten the derivation of new understanding and treatment of disease. Stud Health Technol Inform 2005;112:100–9.

8. NIH Data Sharing Policy and Implementation Guidance. 2003. (Accessed at http://grants.nih.gov/grants/policy/data_sharing/data_sharing_guidance.htm.)

9. Van Horn JD, Ishai A. Mapping the human brain: new insights from FMRI data sharing. Neuroinformatics 2007;5:146–53.

10. Teeters JL, Harris KD, Millman KJ, Olshausen BA, Sommer FT. Data sharing for computational

neuroscience. Neuroinformatics 2008;6(1): 47–55.

11. Fennema-Notestine C, Gamst AC, Quinn BT, et al. Feasibility of multi-site clinical structural neuroimaging studies of aging using legacy data. Neuroinformatics 2007;5:235–45.

12. Butcher J. Alzheimer's researchers open the doors to data sharing. Lancet Neurol 2007;6:480–1.

13. Liu Y, Ascoli GA. Value added by data sharing: long-term potentiation of neuroscience research. A commentary on the 2007 SfN Satellite Symposium on data sharing. Neuroinformatics 2007;5:143–5.

14. Martone ME, Tran J, Wong WW, et al. The cell centered database project: an update on building community resources for managing and sharing 3D imaging data. J Struct Biol 2008;161(3):220–31.

15. Altman M. The clinical data repository: a challenge to medical student education. J Am Med Inform Assoc 2007;14:697–9.

16. BIRN Data Repository. 2008. (Accessed at http://nbirn.net/.)

17. Gridsphere Portal Framework. (Accessed at http://www.gridsphere.org/gridsphere/gridsphere.)

18. Martone ME, Gupta A, Ellisman MH. E-neuroscience: challenges and triumphs in integrating distributed data from molecules to brains. Nat Neurosci 2004;7:467–72.

19. Marcus DS, Olsen TR, Ramaratnam M, Buckner RL. The Extensible Neuroimaging Archive Toolkit: an informatics platform for managing, exploring, and sharing neuroimaging data. Neuroinformatics 2007;5:11–34.

20. Ozyurt IB, Brown GG, Grethe JS, Morphometry BIRN, FIRST BIRN. A general, extensible system for human brain imaging data retrieval and maintenance. In: Annual Meeting of the Organization for Human Brain Mapping. 2004; Budapest, Hungary.

21. Ozyurt IB, Wei D, Keator DB, et al. A user-friendly, web-accessible system for the management, discovery, retrieval, and analysis of clinical and brain imaging data. In: Human Brain Project Neuro-Informatics meeting. 2004; Bethesda, MD.

22. Turner JA, Fennema-Notestine C, Martone ME, et al. You say potato, I say potahto: ontological engineering applied within the Biomedical Informatics Research Network. In: Society for Neuroscience. 2006; Atlanta, GA.

23. Gupta A, Bean C, Bug W, et al. Creating a community-based knowledge management framework for integrating neuroscience data via ontologies. In: Society for Neuroscience. 2006; Atlanta, GA.

24. Gupta A, Astakhov V, Boline J, et al. Data Federation in the Biomedical Informatics Research Network: tools for semantic annotation and query of distributed multiscale brain data. In: Society for Neuroscience. 2007; San Diego, CA.

25. BIRNLex. 2007. (Accessed at http://xwiki.nbirn.net:8080/xwiki/bin/view/+BIRN%2DOTF%2DPublic/About+BIRNLex.)

Chapter 3

Mediator Infrastructure for Information Integration and Semantic Data Integration Environment for Biomedical Research

Jeffrey S. Grethe, Edward Ross, David Little, Brian Sanders, Amarnath Gupta, and Vadim Astakhov

Summary

This paper presents current progress in the development of semantic data integration environment which is a part of the Biomedical Informatics Research Network (BIRN; http://www.nbirn.net) project. BIRN is sponsored by the National Center for Research Resources (NCRR), a component of the National Institutes of Health (NIH). A goal is the development of a cyberinfrastructure for biomedical research that supports advance data acquisition, data storage, data management, data integration, data mining, data visualization, and other computing and information processing services over the Internet. Each participating institution maintains storage of their experimental or computationally derived data. Mediator-based data integration system performs semantic integration over the databases to enable researchers to perform analyses based on larger and broader datasets than would be available from any single institution's data. This paper describes recent revision of the system architecture, implementation, and capabilities of the semantically based data integration environment for BIRN.

Key words: Cyber infrastructure, Neuroinformatics, Biomedical informatics, Distributed information management, Database, Data federation, Semantics, Information integration, Ontology

---

1. Introduction

The Biomedical Informatics Research Network (BIRN) is an initiative within the National Institutes of Health that fosters large-scale collaborations in biomedical science by utilizing the capabilities of the emerging cyber infrastructure (high-speed networks, distributed high-performance computing, and the necessary software and data integration capabilities) *(1)*. Currently,

Vadim Astakhov (ed.), *Biomedical Informatics,* Methods in Molecular Biology, vol. 569
DOI 10.1007/978-1-59745-524-4_3, © Humana Press, a part of Springer Science+Business Media, LLC 2009

the BIRN involves a consortium of more than 30 universities and 40 research groups in the US and UK participating in three testbed projects and associated collaborative projects centered on brain imaging and genetics of human neurological disease and associated animal models. These groups are working on large-scale, cross-institutional imaging studies on Alzheimer's disease, autism, depression, and schizophrenia using structural and functional magnetic resonance imaging (MRI). Others are studying animal models relevant to multiple sclerosis, attention deficit disorder, and Parkinson's disease through MRI, whole brain histology, and high-resolution light and electron microscopy. These testbed projects present practical and immediate requirements for performing large-scale bioinformatics studies and provide a multitude of usage cases for distributed computation and the handling of heterogeneous data in a distributed system. The goal of the data integration and application integration team for the BIRN is to develop a general-purpose semantically enhanced information integration framework that neuroscientists can use in their research (*see* **Fig. 1**). The data to be integrated ranges from 3D volumetric data of nerve components, to image feature data of protein distribution in the brain, to genomic data that characterize the anatomical anomalies of different genetically engineered mouse strains. There are a number of integrated schemas over different combinations of these sources designed for different study groups.

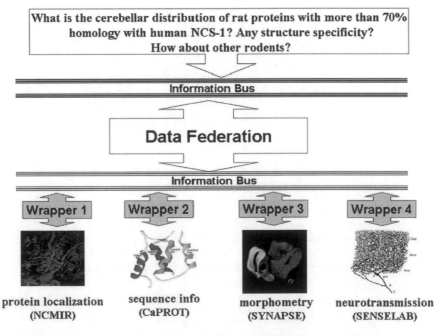

Fig. 1. A typical research question which requires integration over multiple heterogeneous sources.

2. System Architecture

BIRN's data integration environment provides a set of tools and system infrastructure for all phases of the data integration process (i.e., registration of sources to semantic queries) *(2–8)*. This infrastructure **(Fig. 2)** consists of a data integration server; an API for schema registration, view definition, and query building; clients for registration, view definition, and query composition; clients for semantic data mapping and query; a number of domain-specific clients that have incorporated access to the data integration environment (e.g., the Mouse BIRN Atlas Tool); and a set of tree- and graph-structured ontologies and ontology-data mappings. The latter provides the necessary semantic information, tied to external ontologies, so that integrated queries can be executed across the sources. The overall architecture of the system is given in **Fig. 2**.

2.1. Gateway

The point of entry is the data integration environment's (i.e., Mediator's) Gateway. It handles all communication with Mediator clients. The Gateway's principal functions are to: listen on a secure

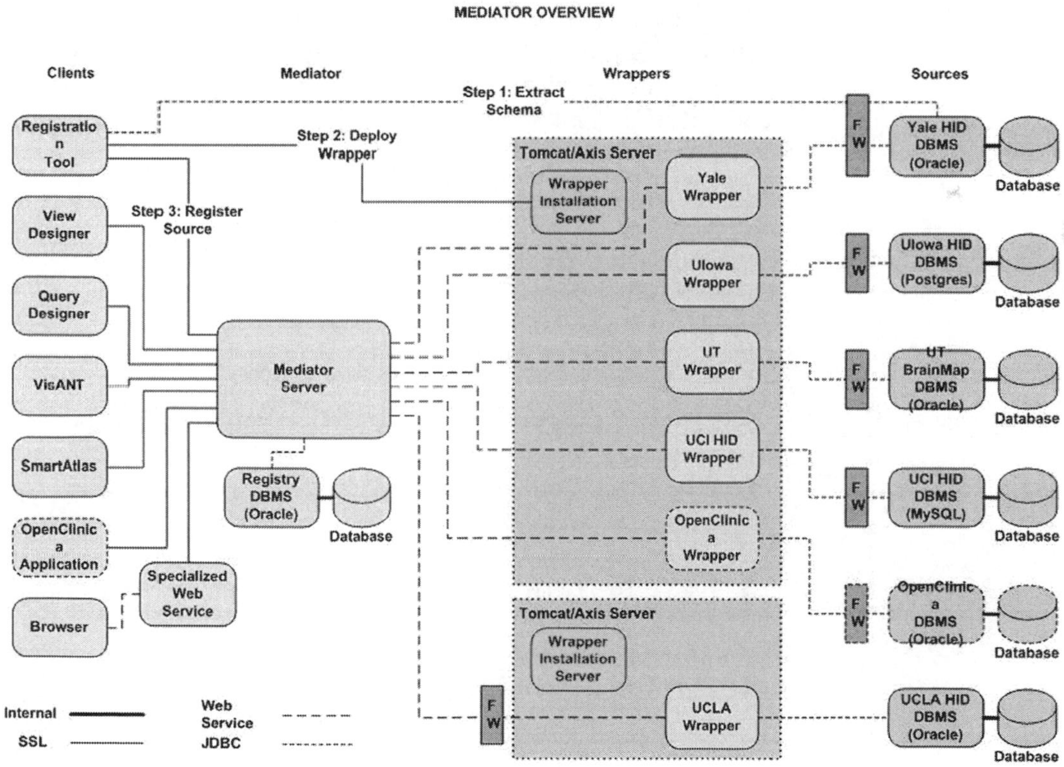

MEDIATOR OVERVIEW

Fig. 2. The general architecture of the Data Integration system.

socket for a client XML request; authorize, validate, and classify the request; process the request; and return an XML response The Gateway reflects typical request–response server design spawning threads for each request.

- Mediator requests fall into one of three categories:

- Registration (update of the Registry)

- Registry query (query against the Registry)

- Data query (query against the data sources)

For a standard data query, the Mediator requests a plan from the Planner and dispatches the plan to the Executor (these modules are described below). For certain types of data queries, such as a keyword query, the Gateway will call on the Preprocessor (**Subheading 4**) to recast the query as a standard data query which can then be handled by the planner and executor.

Access control is a very important aspect of a practical information integration system. Within the Mediator, this is accomplished in two stages – user authentication and the authorization of a user by registered data sources (if required). In order to allow local control of data permissions and policies, the authentication must be handled by the local data sources. To that end the Mediator allows user credentials to be passed to the sources for authentication and authorization.

2.2. Registry

The Registry is the Mediator's internal database of data types, sources, relations, foreign-key relationships, functions, integrated views, users, permission policies, and related data. The relations and functions are source-specific and reflect the portion of a BIRN-participating institution's research data they are sharing with the BIRN and what permission policies they wish to apply to their data. The Registry has an API that is used for registration of data sources and during query planning. The Registry API is based on the DAO (Data Access Object) object-relational mapping. So, the Registry has a relational model that is represented by an object model in the Mediator API.

2.3. Planner

The planner's purpose is to take a Mediator query and produce a logical query plan to hand to the executor. A query plan tells the executor in what order to gather data from the different sources, which data to transfer between what sources, and from which relations or functions to gather the mentioned data. Usually, it is possible to construct several different plans for a particular query where these possible plans may have very different execution characteristics. Therefore the planner attempts to choose, from among possible plans, those with better retrieval times. The planner uses a variety of heuristics to find "good enough" plans, and to find these plans within a reasonable amount of time. This approach is necessary because finding the best of all possible plans, for a particular query, is an NP-complete problem and takes an extraordinary amount of time for all but the most trivial queries.

To this effect, the planner **(Fig. 3)** attempts to find plans that: (a) minimize network traffic, (b) minimize computation (such as joins) at the mediator, and (c) minimize undesirable computations at the sources (such as Cartesian products). Among the strategies employed, in pursuit of these goals, are: (a) pushing as much filtering as possible into all the sources involved, (b) chunking ordered plans into source-specific subplans, (c) reassigning, where possible, specific computations from the mediator to one or more sources, and (d) the applying of heuristics to select those plans with large source-specific chunks.

To produce a plan, for a particular query, the first action of the planner is to construct a tree representation of that query. This tree has, as leaves, all base relations and functions of the query, and, as internal nodes, all the Boolean operators that relate these base components to one another. In a process called unfolding, the planner replaces any integrated view references it finds in the query, by subtrees constructed from the views' definitions.

Unfolding is a query rewriting procedure. A query definition consists of a head and a body. The query head simply specifies the names, data types, and order of the members of the return tuples. The body of a query is an arbitrary Boolean combination of predicates. These predicates are of two types: (a) items which directly represent entities in remote sources, and (b) references to views, the definitions of which are known to the query planner, we will call these ivds (for integrated view definitions). Examples of the former are DBMS tables and views as well as API functions, etc. For our purposes, ivds are definitions which have structure and

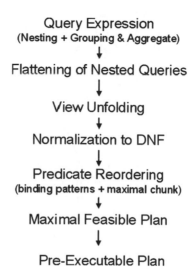

Fig. 3. Planner.

semantics identical to query definitions (i.e., they have a head and a body; the body is a Boolean combination of predicates, etc.). Unfolding is a procedure in which all references to ivds found in the body of a query are replaced by the ivds' respective bodies (with suitable parameter name replacements). It should be noted that due to the fact that ivds may themselves contain references to additional ivds, unfolding must be a recursive process.

This is a recursive process, as integrated views may be nested within each other to any arbitrary depth. Generally speaking, the planner traverses and manipulates this tree in its search for well-behaved plans.

The meaning of the phrase "well-behaved plan" is somewhat difficult to define precisely. For a particular query it may be feasible to construct many different plans. In many cases the "plan times" (plan construction time + plan execution time) of different plans for the same query vary widely. Assuming it were possible to construct a distribution of retrieval times for all the different plans feasible for a query, a well-behaved plan should execute within one order of magnitude of the fastest possible plan (of course this is only our arbitrary notion). However, in the general case, none of the computations required to evaluate a plan by the above criteria can be computed "before the query," which is when we need the information, nor in a time shorter than many multiples of the time it would take the actual plan to execute. In practice, however, this has not proved to be a great problem. There are many well-known heuristics that can be employed to optimize the construction and execution times of query plans. We have used the application of these heuristics as surrogate criteria to determine what a "well-behaved plan" is. For our purposes a "well-behaved plan" is one that was constructed using all the appropriate heuristics.

The next step involves transforming the tree into its equivalent disjunctive normal form (DNF). The result is a two-level tree representing a disjunction of conjunctive queries. This procedure is carried out to simplify the process of finding feasible plans to the query. Assuming that there are no unsafe variables in the head of the query, finding a feasible plan for the original query reduces to finding properly ordered plans for all the constituent conjunctive queries of the DNF. This is a process which is significantly simpler than attempting to order an arbitrary Boolean query. Additionally, once ordered, the DNF allows for easy source-specific chunking within the constituent conjunctive plans.

The ordering algorithm as such is quite simple. It keeps a list of all the relations and functions found in a conjunctive query. Let us refer to these functions and relations as the elements of the query. Each element is considered in turn with the use of a variety of heuristics. As soon as a suitable element is found, it is removed from the list and placed in the next position of the prospective plan, whereupon the process repeats itself with the now smaller list. The heuristics used guarantee that if there is at least one feasible ordering of the conjunctive query, it will be found. The algorithm stops when no more elements can be placed in the plan, either because none remain in the list, or because none of the remaining elements are suitable for placement in the plan. This algorithm does not yield the best possible plan for a query, but it runs in time $O(n^2)$, where n is the number of elements in the conjunctive query, which is quite acceptable for this process.

2.4. Executor

The execution engine translates the preexecutable plane object that it receives from the Planner into the specific set of instructions (XML) and sends it to the wrapper server. The wrapper will convert this XML message to the appropriate language of the source, execute the query at the source and returns the results back. The results retrieved from the wrapper are stored temporarily in the execution engine database to be used following source conjunctions in the plan. To be efficient, only the data that will be needed in the later stages are stored and the rest is ignored. At the end of the execution, the results from each disjunction in the plan are unioned with each other and are sent back to the client. The intermediate values are stored in a relational database (MySQL). The Variable Management Algorithm specifies which variable values are going to be stored at each step of the execution and how the data at the local database will be updated as new data is received from the wrapper. This algorithm uses a Variable Reservoir for the storage of variable values. Intermediate values retrieved from the wrapper are stored in the execution engine database and the information of which variable is stored in which table is located in this Variable Reservoir.

2.5. Wrapper Service

The wrapper is the middleman between the mediator and a source. It accepts the query from the mediator, translates it into source query language, executes it against the source, and finally returns the results back to the Mediator in the Mediator's format. The wrapper server is stand-alone software which acts as a container of wrapper instances. Each source has its own dedicated wrapper instance. The server accepts three kinds of requests: (a) requests to add or remove a wrapper instance, (b) requests to execute a query, and (c) requests to retrieve results. Once the query is executed, the wrapper returns a cursor to the mediator, and the latter uses it to fetch the results in batches allowing for it to handle large results. The mediator sends a request to the wrapper server along with the cursor. The wrapper server figures out which result is needed in the result repository, returns the next batch in the result, and updates the cursor position. In the request, the mediator may include an optional flag to indicate whether it wants the next, previous, first, or last batch. In addition, every query or result has an expiration date so that the query is inter-rupted if it does not finish before the expiration. The result is also deleted after it expires.

3. Integration Framework

3.1. Distributed Heterogeneous Data Resources

A participating institution's data repository, or a *source*, is assumed to be relational with a binding pattern for every disclosed relation. Every schema element (relations and attributes) has a descriptor for keyword search, and a so-called semantic-type that can be used to map the element to ontology. Furthermore, a data source may disclose a set of functions that are internally treated as rela-tions with the binding pattern ($\overline{b}$, f) where $\overline{b}$ represents a set of bound arguments and the single f is the free output variable of the function. All predicates have one or more valid binding patterns. A binding pattern for a particular predicate is nothing more than a list of binding specifiers of the same arity as the predicate. A valid binding pattern is one which allows for the evaluation of the predi-cate. For a specific binding pattern each specifier states whether its positionally corresponding predicate parameter must be bound (input), must be free (output), or can be either (input/output) so that the predicate may be evaluated. Usually tables in RDBMSs have a single binding pattern containing specifiers only of the input/output variety. On the other hand, functions and procedures usually have much more restrictive binding patterns.

In Mouse BIRN, for example, specialized functions are used to compare the distributions of proteins in a set of user-specified regions in the brain. Using this model also enables us to treat computational sources such as the R statistical package as a "data source" that contributes functions but no relations. Integrated views are written using standard data sources as well as these func-tions. We have also designed source-specific wrappers for sources such as Oracle, Oracle Spatial, and PostgreSQL, where a generic query can be translated into the appropriate flavor of SQL and functions supported by the specific sources' systems.

Previous versions of the data integration environment were based on the assumption that users should have knowledge of the semantics of the sources that they are going to use. However, this knowledge is not reasonably available in the case of a large collection of heterogeneous and distributed resources. So the latest version uses common semantic information to navigate and integrate such heterogeneous sources. By using ontological information and data semantics for these data sources an end user can query, inte-grate, and interrelate various sources without having to examine the low-level schema details. In essence this allows researchers to "Query through Ontology."

3.2. Ontologies

Ontology is a term-graph whose nodes represent terms from a domain-specific vocabulary, and edges represent relations that also come from an interpreted vocabulary. The nodes and edges are typed according to a simple, commonly agreed upon set of types produced by testbed scientists. The most common interpretation is given by rules such as the transitivity of *is-a* or *has-a* relations, which can be used to implement inheritance and composition. However, there are also domain-specific rules for relationships such as *volumetric-subpart* (brain-region→brain-region) and *measured-by* (psych-parameter→cognitive-test) that need special rules of inference. For example, if a brain-region participates-in a brain-function (such as "working memory"), and the brain-function is measured-by a cognitive-test, then the cognitive-test functionally-tests the brain-region.

In the current framework, ontologies are represented as a set of relations comprised of a set of nodes and a set of edges, with appropriate attributes for each node and edge. Other operations, including graph functions such as path and descendant finding and inference functions like finding transitive edges, are implemented using an API of functions.

3.3. Mapping Relations

In the current global as view (GAV) setting of the system, the burden of creating proper integrated views over data sources is on the integration specialist who works with the domain scientists to capture the requirements of the application at hand. This often leads to the pragmatic problem that the relationships between attributes disclosed by different sources and between data values are, quite often, not obvious. To account for this, the system has additional *mapping relations*. Currently there are three kinds of mapping relations.

The *ontology-map* relation: maps data-values from a source to a term of a recognized ontology such as BIRNLex, used within BIRN for annotation data sources. The BIRNLex provides terms, utilized by BIRN scientists in the context of their research, covering neuroanatomy, molecular species, behavioral and cognitive processes, subject information, experimental practice and design, and associated elements of primary data provenance required for large-scale data integration across disparate experimental studies. Many of the terms are drawn from existing terminologies and ontologies such as the Unified Medical Language System (UMLS), the Foundational Model of Anatomy (FMA), Neuronames, the Gene Ontology, the Functional Genomics Ontology (FUGO), and the Phenotype and Trait Ontology (PATO). These terms will form the basis of a more formal ontology describing the multiscale investigation of neurological disease.

A *joinable* relation: links attributes from different relations if their data types and semantic types match.

The *value-map* relation: maps a Mediator-supported data value or a Mediator-supported attribute-value pair to the equivalent attribute-value pair disclosed by the source. For example, the Mediator may disclose a demographic attribute called "gender" with values {male, female}, while a source may refer to "sex" with values {0, 1}, while another source, with its own encoding scheme, may call it "kcr_s57" with values {m, f}. We have a preprocessor module that uses a look-up function to make a substitution.

3.4. View Definition and Query Languages

The query and view definition languages are very similar to each other. Both are specified in XML and their syntax overlaps to a large degree. The structure of both languages is based on a subset of Datalog that allows arbitrary Boolean combinations of predicates and aggregation, but disallows recursion. The languages make no restrictions on the types of entities that may be included within definitions so long as they can be represented by predicates. In our case we use predicates to represent relations, functions, and view references. All definitions have a head and a body. The head has a predicate structure and specifies the list of the inputs and outputs of the defined view or query. The structure of the body consists of a Boolean combination of predicates and specifies the flow of data between the represented entities. For example, a query definition may look as follows (in pseudo code for briefness):

$$Q(X,Y) \leftarrow (R(X,W) \text{ or } V(X,W)) \text{ and } F(W,Y)$$

Here Q(X,Y) represents the head of the query and (R(X,W) or V(X,W)) and F(W,Y) represents the body. Within the body, R may be a relation, V a view reference, and F a function possibly with W as input and Y as output.

The definition languages assume the existence of a registry where information such as source addresses, data types, semantic types, view definitions, and binding patterns are stored. The registry information complements all query and view definitions and is essential for the eventual execution of any query.

Query definitions may contain aggregation functions but view definitions may not. We place aggregation functions in the head of the query with the names of the columns to be aggregated as input parameters. Here is an example:

$$Q(X, F(Y)) \leftarrow R1(X, Z) \text{ and } R2(Z, Y)$$

In this query X is a group-by column, Y is the column to be aggregated, and F is the aggregation function. In those cases where an aggregation function can be directly computed at a source, we simply pass an appropriate translation of the containing query to the source for evaluation. Typically, relational sources are capable of evaluating several aggregation functions such as MIN, MAX, AVE, etc.

In those cases where a source cannot compute the aggregation function requested, the mediator assumes that task. Essentially the mediator creates a new query by stripping the aggregation function references from the head of the original query and replaces them with the names of their respective aggregation columns. The mediator then places the results of executing the new query in a temporary table and performs the aggregation on it. In our above example the temporary table would be created from the new query:

$$Q' (X,Y) \leftarrow R1(X,Z) \text{ and } R2(Z,Y)$$

The definition languages also allow safe negations, where all variables in the negated predicates are bound. Nested queries are supported within other queries but not within views. A keyword query capability allows the use of keywords to search the term-index-source to find relations that can be joined *(9)*.

3.5. Optimizations

Several optimization techniques were used. We pushing selects to as many leave nodes of query plan tree as possible. That means that we try to push subqueries to be executed at sources. Also we do not allow cross products within source conjunctions. At the same time we try to minimize cross products between source conjunctions.

Semi joins also used to optimize query execution.

4. Architecture Extensions to Support Semantically Based Integration

To support mapping to ontology, semantically based integration, and the ability to query through ontological concepts, we have developed several architectural extensions. These extensions include a Data annotation tool which helps in generating a map between the data and ontology, a Term-Index-Source that keeps this mapping information, and query preprocessor and postprocessor modules.

4.1. Term-Index-Source

The data integration environment has a specific source called the *term-index-source* that maintains a mapping between data from registered sources and ontological terms defined by various ontologies. This is used by the system to preprocess queries. We have also developed a Concept Mapping client (*see* **Subheading 5**) to assist in the creation of mapping entries in the *term-index-source* at the time of source registration. This tool allows the user to select ontology such as BIRNLex *(12–13)*, browse for relevant concept term identifiers, and associate them with *data objects*.

The client also connects to an arbitrary relational database source using JDBC, and by retrieving the schema metadata, makes it possible for the user to select one or many table-field-value triples to comprise the *data object*. The *data object* is expanded to include related data via primary–foreign key relationships inherent in the metadata. There are also other automation features such as mapping templates to facilitate the mapping of large sources. The result of mapping the ontology to the registered source is a controlled query vocabulary, incorporating ontological terms, that is made available immediately to all search clients that make use of the *term-index-source*.

The *term-index-source* is a relational database and programmatic API which provide mechanisms for mapping distributed data to standard ontological concepts. **Table 1** shows an example of the mapping provided for the Mouse BIRN testbed's Cell Centered Database (CCDB).

At the same time, ontological sources such as BIRNLex and UMLS provide information about how different ontological concepts are linked to one another. That information is used by the Preprocessor to find the shortest path between concepts. **Figure 4** demonstrates how distributed data mapped to two different ontological concepts can be linked using an ontological source like UMLS. Initial data in the ontology mapping taken from TIS and the two UMLS concepts can be linked through the shortest path. It also graphically illustrates the mapping inference.

As shown in the example, mapping of the "Project" table to a UMLS concept leads to the mapping of all table fields to the same concept; in contrast, single value mapping does not lead to any inference. Figure 4 also demonstrates that two concepts can be connected through various paths.

We have implemented several algorithms to support relevant data retrieval using ontology concept to concept paths, ontology to data mapping, PK–FK constraints, and ontological data joins that are based on concept ids rather than actual data. We have developed the concept of a data object where data grouped into the object is marked by an ontological concept.

Table 1
Example term-index-source view

Ontol source	Ontol ID	DB name	Table name	Field name	Value	Type
UMLS	C27853	UCI	exper	Test	L	str
UMLS	C96368	Barlo	ene	Name	Daf2	str
UMLS	C25914	Barlo	expres	Sign	123	int

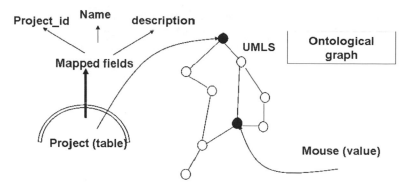

Fig. 4. Data to ontology mapping.

4.2. Automatic Query Generation Algorithm

4.2.1.

At the first step, a user is expected to provide a list of key words defining areas of interest. The Term Index Source (TIS) engine takes the list and performs a key word search over predefined ontological sources and returns a list of ontological concepts. The user reviews the concept list and picks the subset that reflects the areas of interest.

4.2.2.

Next, the TIS engine uses this subset of terms to extract a list of sources and relations from the maps stored in the TIS. Sources are ranked based on concepts mapped to the source data. Sources which have data mapped for each concept in the list will be ranked as top candidates. For each of these candidate sources, the engine extracts a list of mapped tables/fields. The algorithm starts from fields mapped to the concepts and creates a minimal spanning tree constructed to generate a source local view based on PK–FK constraints. For each source s in these sources:

For each pair of relations that are mapped in the *term-index-source* <u, v> in s we compute the path of the foreign key–primary key between them from the database constraints: if such a path exists, then for each relation-column in the fk–pk path we add an edge to the directed graph g, then compute the shortest path sp on g for <u, v> using the Dijkstra algorithm, create a view v representing the join across each relation-column in sp, and map the view to the concept ID associated with s. Such views assemble all data relevant to provided ontological concepts. The PK–FK constraints are used to join mapped tables directly or through intermediate tables. Intermediate tables are not included in the final view.

4.2.3.

Lastly, the user can request data from individual views generated at each candidate source or it can request data to be returned from a join across all those views. This join is done based on the ontological markup rather than string matching. For example, the value "caudate putamen" may be joined with "neurostriatum" if they are both marked as being the same concept id.

4.3. Data Pointer/URL Concept

Some sources have data associated with a concept but the data may actually be stored in a separate table. We have introduced the "Data_URL" concept that represents this associated data. A source provider is required to map all associated data to a "Data_URL" concept to allow them to be automatically discovered. This is critical for researchers performing queries who wish to find actual datasets.

4.4. Preprocessor

The Preprocessor utilizes developed algorithms for semantic keyword queries. It uses an *automatic query generation algorithm* in which an integrated view is automatically generated during preprocessing. The preprocessor uses the *term-index-source* to find relations mapped to those ontological ids. An integrated view can join relations through an attribute mapped to the same concept for both relations that also have the same data type. In the case where keywords are mapped to different concepts, we are looking for the shortest path between the two concepts and then looking for associated relations to support this indirect join. If relations are from the same source, then they will be joined based on the source's foreign key constraints which are recorded in the Registry during source registration. A minimal spanning tree algorithm is used to find a minimal graph to join relations. **Figure 5** presents

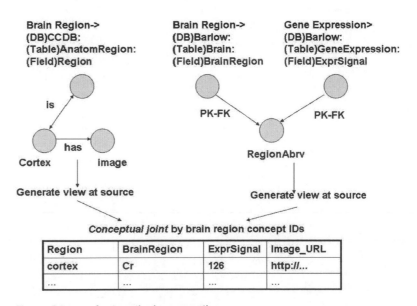

Fig. 5.   Process of automatic view generation.

an example of the process of automatic view generation that is required to answer a key-word-based query with two key words: "Brain region" and "Gene expression." Those key words are mapped to appropriate ontological concepts ids.

The figure illustrates that "Brain region" and "Gene expression" concepts can be linked through the primary key foreign key constraint at the source. Then, the local views can be unioned or joined based on concept IDs extracted from the ontological source. Based on mappings provided by the term-index-source, locations of the data relevant to the explored concept can be found. Finally, SQL queries are issued to collect actual data.

4.5. Postprocessor

The postprocessor prepares output data in formats applicable for various BIRN testbeds and client tools.

5. Mediator Clients and User interaction

The data integration environment has available tools that provide end-to-end functionality, that is, the ability to take a new source, register/map it, and issue semantic queries against the source. First, in order to support basic data integration activities (e.g., source registration) there are several clients that are available.

5.1. Database Registration Tool

This provides a graphical user interface (GUI) for source registration. Registration is the process by which BIRN-participating institutions disclose the data to be shared across the BIRN. This tool (**Fig. 6**) supports multiple database systems and it currently supports Oracle, MySQL, PostgreSQL, and SQL-Server.

5.2. Concept Mapping Tool

First, a researcher familiar with data in a source must map this data to concepts in an ontology (see Term-Index-Source earlier). The Concept Mapping tool is software written in Java comprising a graphical user interface written using the Swing Framework and database client logic using the JDBC library. This ontology-based annotation tool aids the subject matter expert in the construction and editing of a controlled vocabulary metadata for a database. The controlled vocabulary is a list of ontological concepts, which are carefully selected from the BIRNLex ontology. These concepts are used to tag units of information to improve the accuracy of the search process as well as to effectively publish a catalog of the database contents.

This approach improves the accuracy of query formulation by addressing the ambiguity of natural language, thus reducing the number of irrelevant results and dramatically increasing the precision performance of a database query. Similarly, the recall

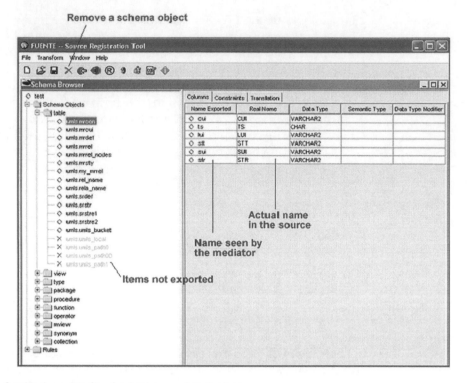

Fig. 6.   Graphical user interface for database registration tool.

performance is improved by term expansion using interterm relationships in the ontology graph.

To map the data, the tables in the database are displayed in a list (**Fig. 7**) and the user can drill down to the primary data fields. By double clicking on the field a popup dialog displays all field values for the selected field in another list. If there is related data discovered through the foreign key–primary key relationships specified in the RDBMS metadata, those tables and fields are listed in the related data panel whose field values can be explored in the same manner as the primary data.

When the user double clicks on the field value, the actions specified by the checkboxes will be taken. (a) Map the value or (b) Map All Values and optionally retrieve all related data fields by fk–pk chains. When the user is finished with mapping the fields for a given table, the "Commit Mapping" button is pressed to create a new Data Object from these field-value pairs and tag it with the working concept term selected in the Ontology Pane. Double clicking on a mapping in the Mappings list displays a popup dialog containing the mapping in a Data Object representation. This allows the individual elements of the Data Object to be edited. The user can also test the query to the database source using this Data Object by pressing the "Preview Query" button. This populates a data table with the query results under the "Data" tab in

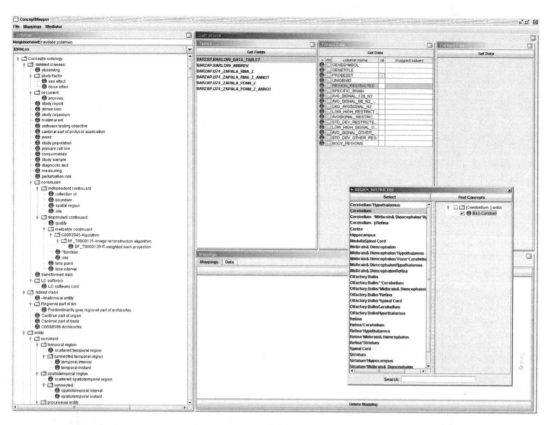

Fig. 7. "Concept Mapper." *Left screen* shows ontology (from .owl file). *Right panels* show a source schema which is under mapping process.

the "Mappings" panel. Testing the query verifies that the data is still available from the source and that it conforms to the original intent of the mapping to the given concept term.

By selecting "Run distributed query" the user can select a list of project files to be included in a "lightweight" mediated query. The popup dialog allows the user to issue a query using concept terms, and this feature searches all mapped sources to these concepts contained in the selected projects. Using the mapped Data Objects, queries are issued to the mapped sources and the results are joined into a single table for viewing. This is a high-performance query because having the mapping table obviates the need for query planning, and the set of all relevant sources is known at query time.

Other GUI clients allow the testing and query of registered sources: the View Designer, which allows a user to develop integrated views across resources, and the Query Designer, which allows a user to submit queries. The following **Fig. 8** demonstrates these interfaces. On the top screen, gene expression data for various brain regions were joined with information about

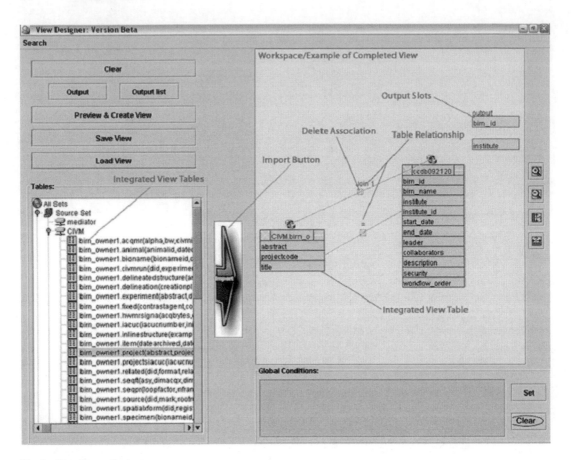

Fig. 8.   View/Query Designer.

brain regions. On the bottom screen, the resulting view was queried and the result provided in the table on the right.

These tools require some knowledge of the data sources, which is appropriate for data source administrators who are registering their data source or data integration engineers who wish to build specific multisource queries. To support a researcher who might not be familiar with details of registered sources, a *Unified Data Integration Workspace* was proposed that helps browse distributed database on semantic information associated to the sources **(Fig. 9)**.

This workspace integrates tools for ontology markup, ontology browsing, and query design such as: MBAT, SmartAtlas, VisANT-Builder, and others developed at BIRN (accessible through BIRN portal: www.nbirn.net).

5.3. WebService API

We developed web services-based APIs to answer multisource queries through the mediator (http://bcc-stage-04.nbirn.net:8080/axis/Query.jws). These APIs are built to satisfy core functions of the

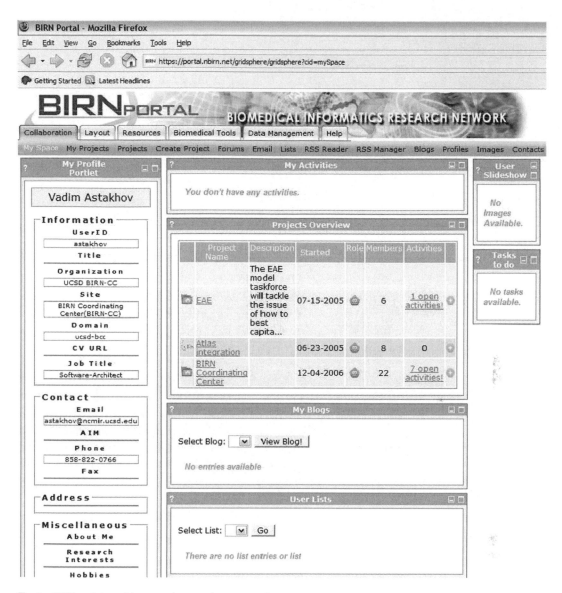

Fig. 9.   BIRN portal provides a work space for a researcher.

data integration environment and also specific requests from participating testbed users. At the interface level we are working to provide methods to submit queries in languages (XCEDE *(14)*, SQL) other than the mediator's or to generate them from query-building clients. Also at the interface level, we provide for output formats (XCEDE) different from the mediator's native one. At the processing level these services generate native mediator queries through the use of parametrizable templates or through straight translation from languages such as SQL.

6. Current Work and Conclusions

A key goal for our work is providing biomedical researchers, who do not have any expertise in ontologies or query formulation, the ability to easily query a distributed collection of data resources. The researcher will not know all the particulars of each data source or what data is contained in them. Therefore, we have been building this system for researchers to be able to easily perform key word type queries. This system allows researchers to navigate through hierarchically organized ontological-concepts which are mapped to actual data. The main benefit of this approach is that it lets us abstract the details from the user so that they do not need to know the detail of sources and be an expert in query formulation, which they are not. The process of using an ontology based query vocabulary provides the necessary framework for the user relating to the nature and extent of data available from the federated database. We continue to work on the ability to formulate a natural language query like "what are the URLs of confocal microscopy images of brain regions related to the alpha-synuclein protein that have gene expression data."

The Mediator team is also working on strategies to better handle spatial data related to image information. In the Mediator we are developing indexes of the extent of every spatially registered image instance, coverage constraints in spatial coordinates, and articulation mapping with ontology nodes. We also need to extend the environment to produce more efficient plans and perform more judicious source selection.

7. Notes

1. Cyber infrastructure was developed for the Biomedical Informatics Research Network www.nbirn.net.

2. Data integration platform provides an environment for semantic integration of the data which are distributed across various scientific groups.

3. Ontology and Mediator Server represent the core of the semantic integration where heterogeneous data mapped to ontology and mediated by "Mediator" server.

4. Developed technologies were adopted by variety different projects such as: Neuroscience Information Framework, National Database for Autism Research, and Community Cyberinfrastructure for Advanced Marine Microbial Ecology Research and Analysis.

Acknowledgment

This work is supported by NIH BIRN-CC Award No. U24 RR019701-01 (NCRR BIRN Coordinating Center Project BIRN005). The authors thank their colleagues and students involved in the BIRN project for their contributions, in particular, Willy WaiHo Wong, Joshua Tran, Asif Memon, and Ilya Zaslavsky.

References

1. J.S. Grethe, C. Baru, A. Gupta, M. James, B. Ludaescher, M.E. Martone, P.M. Papadopoulos, S.T. Peltier, A. Tajasekar, S. Santini, I.N. Zaslavsky, M.H. Ellisman, (2005) "Biomedical Informatics Research Network: Building a national collaboratory to hasten the derivation of new understanding and treatment of disease", Stud Health Technol Inform, 112:100–109, 2005.

2. B. Ludäscher, A. Gupta, M.E. Martone, "Model-based mediation with domain maps", 17th International Conference on Data Engineering (ICDE), Heidelberg, Germany, IEEE Computer Society, 2001.

3. A. Gupta, B. Ludäscher, M.E. Martone, "Registering scientific information sources for semantic mediation", 21st International Conference on Conceptual Modeling, (ER), Tampere, Finland, pp. 182–198, October 2002.

4. M.E. Martone, A. Gupta, B. Ludascher, I. Zaslavsky, M.H. Ellisman, "Federation of brain data through knowledge-guided mediation", In R. Kotter (ed.), Neuroscience Databases: A Practical Guide, Boston, MA: Kluwer, pp. 275–292, October 2002.

5. A. Gupta, B. Ludäscher, J.S. Grethe, M.E. Martone, "Towards a formalization of disease-specific ontologies for neuroinformatics", Neural Netw, 16(9):1277–1292, 2003.

6. M.E. Martone, A. Gupta, M.H. Ellisman, "e-Neuroscience: Challenges and triumphs in integrating distributed data from molecules to brains", Nat Neurosci, 7(5):467–472, 2004.

7. A. Gupta, B. Ludäscher, M.E. Martone, A. Rajasekar, E. Ross, X. Qian, S. Santini, H. He, I. Zaslavsky, "BIRN-M: A semantic mediator for solving real-world neuroscience problems", ACM Conference on Management of Data (SIGMOD) (Demonstration), p. 678, June 2003.

8. V. Astakhov, A. Gupta, S. Santini, J.S. Grethe, "Data Integration in the Biomedical Informatics Research Network (BIRN)", Data Integration in the Life Sciences, Springer, pp. 317–320, 2005.

9. V. Astakhov, A. Gupta, J.S. Grethe, E. Ross, D. Little, A. Yilmaz, X. Qian, S. Santini, M.E. Martone, M.H. Ellisman, "Semantically Based Data Integration Environment for Biomedical Research", 19th IEEE International Symposium on Computer-Based Medical Systems (CBMS 2006), June 2006, Salt Lake City, UT.

10. A. Prince, P. Smolensky, "Optimality: From neural networks to universal grammar", Science, 275(5306):1604–1610, 1997.

11. Tony. A. Plate, "Holographic reduced representation", CSLI Lecture Notes Number 150, 1997.

12. A. Gupta, C. Bean, W. Bug, C. Fennema-Notestine, M.E. Martone, J.A. Turner, J.S. Grethe, "Creating a community-based knowledge management framework for integrating neuroscience data via ontologies", Society for Neuroscience Annual Meeting, 2006.

13. J.A. Turner, C. Fennema-Notestine, M.E. Martone, A.R. Laird, J.S. Grethe, W. Bug, A. Gupta, C. Bean, "You say potato, I say potahto: Ontological engineering applied within the Biomedical Informatics Research Network", Society for Neuroscience Annual Meeting, 2006.

14. D. Keator, S. Gadde, J.S. Grethe, D. Taylor, FIRST BIRN, S.A. Potkin, "General XML schema and associated SPM toolbox for storage and retrieval of neuro-imaging results and anatomical labels", Neuroinformatics, 4(2):199–212, 2006.

Chapter 4

System Biology of Gene Regulation

Michael Baitaluk

Summary

A famous joke story that exhibits the traditionally awkward alliance between theory and experiment and showing the differences between experimental biologists and theoretical modelers is when a University sends a biologist, a mathematician, a physicist, and a computer scientist to a walking trip in an attempt to stimulate interdisciplinary research. During a break, they watch a cow in a field nearby and the leader of the group asks, "I wonder how one could decide on the size of a cow?" Since a cow is a biological object, the biologist responded first: "I have seen many cows in this area and know it is a big cow." The mathematician argued, "The true volume is determined by integrating the mathematical function that describes the outer surface of the cow's body." The physicist suggested: "Let's assume the cow is a sphere...." Finally the computer scientist became nervous and said that he didn't bring his computer because there is no Internet connection up there on the hill.

In this humorous but explanatory story suggestions proposed by theorists can be taken to reflect the view of many experimental biologists that computer scientists and theorists are too far removed from biological reality and therefore their theories and approaches are not of much immediate usefulness. Conversely, the statement of the biologist mirrors the view of many traditional theoretical and computational scientists that biological experiments are for the most part simply descriptive, lack rigor, and that much of the resulting biological data are of questionable functional relevance.

One of the goals of current biology as a multidisciplinary science is to bring people from different scientific areas together on the same "hill" and teach them to speak the same "language." In fact, of course, when presenting their data, most experimentalist biologists do provide an interpretation and explanation for the results, and many theorists/computer scientists aim to answer (or at least to fully describe) questions of biological relevance. Thus systems biology could be treated as such a socioscientific phenomenon and a new approach to both experiments and theory that is defined by the strategy of pursuing integration of complex data about the interactions in biological systems from diverse experimental sources using interdisciplinary tools and personnel.

Key words: System biology, Gene expression, Gene regulation, Network topology, Graph analysis, Pathways

Vadim Astakhov (ed.), *Biomedical Informatics,* Methods in Molecular Biology, vol. 569
DOI 10.1007/978-1-59745-524-4_4, © Humana Press, a part of Springer Science+Business Media, LLC 2009

1. Introduction

Availability of large-scale experimental data for cell biology makes it possible to use data integration and systems biology methods to systematically model the behavior of cellular networks and better understand the causal basis of disease and the mechanisms of how it affects an entire organism. Whether we are studying cancer, a neurodegenerative disease like Parkinson's or Alzheimer's, a metabolic disease like diabetes, or the malfunction of the immune system, modern scientists try to understand physiological-level phenomena at the level of a single cell (or a group of cells) – a building block of living organisms. The understanding of the life cycle of a single cell and mechanisms of cell communications are of vital importance for general understanding of living systems.

For example, a cancerous tumor is an uncontrolled proliferation of cells, understanding the causal basis of which needs understanding of functions of these cells and the core mechanisms of gene regulation and gene expression which are described in Section 2. Some of the key questions in gene regulation that are important for various micro- and macroscopic processes are: What genes are expressed in a certain cell at a certain time? How does gene expression differ from cell to cell in a multicellular organism? Which proteins act as transcription factors, i.e., are important in regulating gene expression?

Nearly all of the cells of a multicellular organism contain the same DNA that yields a large number of different cell types. The fundamental difference between a neuron and a liver cell, for example, is which genes are expressed. Furthermore, understanding the gene expression mechanisms may give important clues about various diseases. Some diseases, such as sickle-cell anemia and cystic fibrosis, are caused by defects in single genes; others, such as certain cancers, are caused when the cellular control circuitry malfunctions. An understanding of these diseases will involve pathways of multiple interacting gene products. Here the concepts of molecular networks (i.e., *metabolic pathways, gene regulatory* and *protein interaction networks*) are used to identify groups of molecules that interact in a specific way so as to realize the functions of cells. Many drugs are designed to influence the intracell circuitry and intercell communication, to suppress or stimulate the cell's behavior in one way or another. Therefore full understanding of cell functions, its growth, differentiation, proliferation, and apoptosis (programmed death) requires the understanding of concepts of molecular networks operation, their dynamics and evolution. Section 3 surveys the recent advances in the field of graph-driven methods for analyzing complex cellular networks. The methods are outlined on three levels of increasing complexity, ranging from methods that can characterize global or local structural properties of networks to methods that can detect groups of interconnected nodes, called motifs or clusters, potentially involved in common elementary biological functions.

In Section 4 are summarized recent approaches to data integration and network inference through graph-based formalisms. The goal of this section is to describe information integration methods as well graph-theoretic methods integrating molecular interaction networks and ontologies, and to show how they can be used by biologists to address biologically relevant questions about patterns of protein–protein and protein–DNA interactions.

Section 5 describes the use of ontologies in current biotechnology as a framework for systematizing the onslaught of information encountered in genomics, transcriptomics, proteomics, and other fields of modern biology. Distilling biological knowledge is primarily focused on unveiling the fundamental hidden structure as well as the grammatical and semantic rules behind the inherently related "*omics*" data within the boundary of a biological organism. Sharing vocabulary constitutes only the first step toward information retrieval and knowledge discovery. Ontology is a precise formulation of the concepts that form the basis for communication within a specific field. Once data have been represented in terms of ontology, it is often necessary to transform the data into other representations which can serve very different purposes. This is especially important for bioinformatics and systems biology because of the high degree of heterogeneity of both the format and the data models of the myriads of existing databases. In this section we will provide an introductory analysis of ontologies to bioinformaticists, computer scientists, and other biomedical researchers who have intensive interests in exploring the meaning of the gigantic amounts of data generated by high-throughput technologies.

Finally, we highlight some challenges in the field and offer our personal view of the key future trends and developments in graph-based analysis of large-scale datasets.

2. Gene Regulation and the Systems Biology Approach

2.1. Gene Expression and Gene Regulation

The biological process of gene expression is a rich and complex set of events that leads from DNA through many intermediates to functioning proteins (**Fig. 1**). The process starts with transcription, a complicated set of events that leads from DNA to messenger RNA. A gene's regulatory region contains binding sites for transcription factors (TFs), which act by binding to the DNA (directly or with other transcription factors in a small complex) and affecting the initiation of transcription. In simple prokaryotes, the regulatory

CONTROL OF GENE/mRNA/PROTEIN ACTIVITY

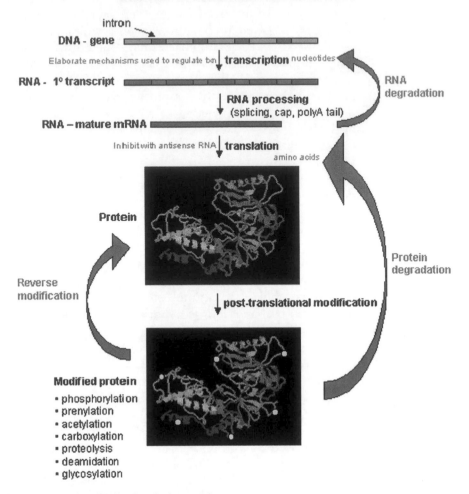

Fig. 1. Information flow from DNA to functioning proteins.

region is typically short (10–100 bases) and contains binding sites for a small number of TFs. In eukaryotes, the regulatory region can be very long (up to 10,000 or 100,000 bases), and contains binding sites for multiple TFs. TFs may act either positively or negatively; that is, an increase in the amount of TF may lead to either more or less gene expression, respectively. Another input mechanism is phosphorylation or dephosphorylation of a bound TF by other proteins.

2.1.1. Transcriptional Modules

Typically, TFs do not bind singly, but in complexes. TFs work by binding to the DNA and affecting the rate of transcription initiation. However, severe complications ensue when interactions with other TFs in a large transcription complex become important. For simple prokaryotes, it is sometimes possible to write out all possible binding states of the DNA, and to measure binding constants for each such state. For more complicated eukaryotes, it is not. It has been hypothesized that TFs have three main functions. Some are active in certain cells and not in others and provide positional control; others are active at certain times and provide temporal control; still others are present in response to certain extracellular signals.

Many binding sites occur in spatial and functional clusters called *enhancers, promoter elements,* or *regulatory modules.* For example, the 480 base pair *even-skipped* (*eve*) stripe 2 "minimal stripe element" in *Drosophila (1)* has five activating binding sites for the *bicoid (bcd)* TF and one for *hunchback (hb)*. It also has three repressive binding sites for each of *giant (gt)* and *Kruppel (Kr)*. The minimal stripe element acts as a "module" and suffices to produce the expression of *eve* in stripe 2 (out of seven *eve* stripes in the developing *Drosophila* embryo). Similar modules for stripes 3 and 7, if they can be properly defined *(2)*, would be less tightly clustered. These promoter regions or modules suggest a hierarchical or modular style of modeling the transcription complex and hence single gene expression, such as that provided by Yuh et al. *(3)* for *Endo16* in sea urchin.

There may be extensive cooperativity between binding sites, even in prokaryotes; for example, one dimer may bind at one site and interact with a second dimer at a second site. If there were no cooperativity, the binding at the two sites would be independent; cooperativity tends to stabilize the doubly bound state. Competition is also possible, particularly for two different TFs binding at nearby sites.

Eukaryotic promoters may have large numbers of binding sites occurring in more or less clustered ways, whereas prokaryotes typically have a much smaller number of binding sites. For N binding sites, a full equilibrium statistical mechanics treatment (possibly oversimplified) will have at least $2N$ terms in the partition function, one for each combination of bound and unbound conditions at all binding sites. The most advantageous way to simplify this partition function is not known because there are many possible interactions between elements of the transcription complex (some of which bind directly to DNA, some of which bind to each other). In the absence of all such interactions, the partition function could be a simple product of N independent two-term factors, or perhaps one such sum for each global active and inactive state.

A further complication is found in the "specific" TFs such as TFIID that assemble at the TATA sequence of eukaryotic transcription complexes, building a complex with RNA polymerase II, which permits the latter to associate with DNA and start transcribing the coding sequence. Finally, signal transduction (e.g., by MAP kinase cascades *(4)*) may act by phosphorylating constitutively bound TFs, converting a repressive TF into an enhancing one.

2.1.2. Exons and Introns, Splicing

In prokaryotes, the coding region is contiguous, but in eukaryotes the coding region is typically split up into several parts. Each of these coding parts is called an *exon*, and the parts in between the exons are called *introns*. Recall that in eukaryotes transcription occurs in the nucleus. For both types of organisms, translation occurs in the cytosol. Between transcription and translation, in eukaryotes, the mRNA must be moved physically from inside the nucleus to outside. As part of this process, the introns are edited out, which is called *splicing*. In some cases, there are alternative splicings, that is, the same stretch of DNA can be edited in different ways to form different proteins. At the end of the splicing process, or directly after the transcription process in prokaryotes, the mRNA is in the cytosol and ready to be translated.

2.1.3. Translation

In the cytosol, mRNA binds to ribosomes, complex macromolecules whose function is to create proteins. A ribosome moves along the mRNA three bases at a time, and each three-base combination, or codon, is translated into one of the 20 amino acids. As with transcription, the rate of translation varies according to experimental conditions *(5)*.

2.1.4. Posttranslational Modification

The function of the ribosome is to copy the one-dimensional structure of mRNA into a one-dimensional sequence of amino acids. As it does this, the one-dimensional sequence of amino acids folds up into a final three-dimensional protein structure. A protein may fold by itself or it may require the assistance of other proteins, called *chaperones*. As previously mentioned, the process of protein folding is currently the subject of a large amount of computational work; we do not discuss it further here.

Degradation DNA is a stable molecule, but mRNA and proteins are constantly being degraded by cellular machinery and recycled. Specifically, mRNA is degraded by a ribonuclease (RNase), which competes with ribosomes to bind to mRNA. If a ribosome binds, the mRNA will be translated; if the RNase binds, the mRNA will be degraded.

Proteins are degraded by cellular machinery, including proteasomes signaled by ubiquitin tagging. Protein degradation is regulated by a variety of more specific enzymes (which may differ from one protein target to another). For multimers, the monomer and multimer forms may be degraded at different rates.

Other Mechanisms Eukaryotic DNA is packaged by complexing with histones and other chromosomal proteins into chromatin. The structure of chromatin includes multiple levels of physical organization such as DNA winding around nucleosomes consisting of histone octamers. Transcriptionally active DNA may require important alterations to its physical organization, such as selective uncoiling. Appropriately incorporating this kind of organization, and other complications we have omitted, will pose further challenges to the modeling of gene regulation networks.

2.1.5. Feedback and Gene Circuits

We have considered some of the mechanisms by which transcription of a single gene is regulated. A key point we have avoided is feedback. Simply stated, the TFs are themselves subject to regulation. This leads to interconnected systems that are more difficult to analyze than feedforward systems. For single-variable systems, there are two major kinds of feedback – positive and negative; for multivariable systems, feedback is more complicated.

Negative feedback is the way a thermostat works: when a room gets too hot, the cooling system kicks in and cools it down; when a room gets too cool, the heater kicks in and warms it up. This leads to stabilization about a fixed point. More complicated negative feedback is also possible, which leads to better control.

Positive feedback can create amplification, decisions, and memory. Suppose your thermostat were wired backward, in the sense that if the room got too hot, the heater would turn on. This would make it even hotter, so the heater would turn on even more, etc., and soon your room would be an oven. On the other hand, if your room got too cold, the air conditioner would kick in, and cool it down even more. Thus, positive feedback would amplify the initial conditions – a small hot temperature would lead to maximum heat, a small cold temperature would lead to maximum cooling. This results in two stable fixed points as final states – very hot and very cold – and a decision between them. Roughly, this is how it is possible for a cell to pick one of several alternative fates and to remember its decision amidst thermal noise and stochastic environmental input.

2.2. Systems Biology Approach

At this point, we have summarized the flow of information from TFs to DNA, DNA to mRNA, mRNA to a protein, and protein back to DNA. Over the years, considerable information has accumulated on the regulatory interactions between TFs and their target genes, proteins, and their modification states in various model organisms such as *Escherichia coli*, *Saccharomyces cerevisiae*, *Caenorhabditis elegans*, *Mus musculus*, *Homo sapiens*, and others. This has allowed the representation of this information in the form of a directed graph, which is commonly referred to as the transcriptional regulatory and protein interaction networks, and signaling and metabolic pathways. Mathematical graphs are a straightforward way to represent this information, and graph-based models can exploit the structure of the transcriptional, translational, and posttranslational regulation, both at local and global levels of organization. Most current research activities concern the dissection of networks into functional modules, a principal approach attempting to bridge the gap between our very detailed understanding of network components in isolation and the "emergent" behavior of the network as a whole, which is frequently the phenotype of interest on a cellular level. Approaches developed for DNA and protein sequence analysis, such as multiple alignment and statistical overrepresentation of parts, are being carried over to address these problems.

The most noted pathways source KEGG *(6)* is centered on enzymatic activities in cellular processes. The KEGG ontology is organized around the concept of binary relations *(7)*, defining relationships between database objects (such as the relationship between reactions, substrates, and products; that between an enzyme and its location in the metabolic pathway; or that between an enzyme and a protein superfamily to which it belongs). Karp et al. (2000, 2002), in BioCyc, on the other hand, define different types of molecules each with its own class, and consider different states of a molecule as different members within a class. Reactions are defined to be independent entities, and distinct relations, called slots, link molecules to the reactions. Each molecule may optionally be tagged with a cellular compartment. Their ontology makes use of the "pathway" concept to define summary abstractions, used for defining data at varying levels of detail. However, like KEGG, the BioCyc system implements a specific data model for its own application. The PathDB *(8)* is a relational database developed for metabolic networks. Here the central element is a "biochem" (e.g., RNA, DNA, Compound) which is used to build other "biochem" objects. The transition is modeled by the explicit representation of a biochemical reaction whose substrates, products, and mediators with their respective kinetic properties are recorded. The Pathways database system *(9)* models pathways as a directed hypergraph where nodes represent pathway elements (substrates and products of a reaction). Pathways DB supports queries where operations such as shortest path, unions and intersections of paths, and node-neighborhoods can be performed. Since Pathways does not present a query language, the exact query capability of the system is unclear. Other works include those of Ochs et al. *(10)*, who developed a metabolic map from a relational model of biochemical interactions, and of Bhalla *(11)*, where a database of chemical reactions is mapped to a system of pathway graphs.

In contrast to KEGG or BioCyc, PathSys *(12, 13)* is based on a generic graph model that can integrate any combination of graph data sources. Consequently it represents a wider range of data types and relationships and can be extended by including any new data source or ontology. Unlike any previous system, PathSys is a general-purpose graph warehouse with its own data definition and query language, augmented with biological data types, and hence can implement any specific graph-structured biological model. The benefit of having an integration platform such as PathSys is that it can be constructed over those databases that typically focus on specific interaction studies [ref], as well as those of process-specific databases such as BioCyc and KEGG focusing on specific biological processes.

Several studies suggest that the transcriptional and protein networks inferred from model organisms may be approximated by a scale-free topology, which in turn implies the presence of a relatively small group of highly connected regulators (hubs or global regulators). Different graph-theoretic principles have been applied to infer various properties of such networks, as well as investigation of the evolutionary principles of the transcriptional regulatory networks across diverse organisms. It is observed that target genes show a much higher level of conservation than their transcriptional regulators. This in turn suggested that the same set of functions could be differently controlled across diverse organisms, contributing significantly to their adaptive radiations. In particular, at the local level of network structure, organism-specific optimization of the transcription network has evolved primarily via tinkering of individual regulatory interactions rather than whole scale reuse or deletion of network motifs (local structure).

Systematic study of complex interaction networks in biological systems described above forms a multidisciplinary branch of modern biology called systems biology. Systems biology can be considered from a number of different aspects. Some sources discuss systems biology as a field of study, particularly, the study of the interactions between the components of biological systems, and how these interactions give rise to the function and behavior of that system (for example, the enzymes and metabolites in a metabolic pathway). Other sources consider systems biology as a paradigm, usually defined in antithesis to the so-called reductionist paradigm, although fully consistent with the scientific method. The distinction between the two paradigms is referred to in these quotation:

"The reductionist approach has successfully identified most of the components and many of the interactions but, unfortunately, offers no convincing concepts or methods to understand how system properties emerge...the pluralism of causes and effects in biological networks is better addressed by observing, through quantitative measures, multiple components simultaneously and by rigorous data integration with mathematical models," *Science*.

Still other sources view systems biology in terms of the operational protocols used for performing research, namely, a cycle composed of theory, computational modeling to propose specific testable hypotheses about a biological system, experimental validation, and then using the newly acquired quantitative description of cells or cell processes to refine the computational model or theory.

3. Graph-Theoretic Methods for Analyzing Biological Data

The availability of molecular interaction graphs has paved the way for new kinds of analyses to systematically investigate the topology and function of these networks using well-understood graph-theoretical concepts that can be used to predict the structural and dynamical properties of the underlying network. The study of network topologies of specific networks in Lukashin et al. *(14)*, or the identification of graph-theoretic properties of these networks in Klamt and Gilles *(15)*, Wuchty et al. *(16)*, and Han et al. *(17)* suggest a biological hypothesis regarding, for instance, unexplored new interactions of the global network or the function of individual cellular components that are testable with subsequent experimentation. Even a simplistic dynamical system originating from small Boolean network models, where nodes represent discrete biological entities (i.e., mRNA or protein), that can be thought to be either on or off, and edges their Boolean relationships ("genotypes"), can give rise to a multitude of designable dynamical outputs ("phenotypes"). Mathematical modeling also enables an iterative process of network reconstruction, where model simulations and predictions are closely coupled with new experiments chosen systematically to maximize their information content for subsequent model adjustments, providing increasingly more accurate descriptions of the network properties. The topological relations underlying graph-based methods can also convey structure to putative pathways. This helps avoiding approaches that test many known sets of molecules without causal interactions [ref]. Furthermore, graph formalisms may provide powerful tools for "omics" data integration to address fundamental biological questions at the systems level.

3.1. Network Topology and Graph Analysis

Global network measures, such as the degree distribution of a node (number of edges it participates in) and the clustering coefficient (the number of edges connecting the neighbors of the node divided by the maximum number of such edges), are the most general characteristics of cellular network analysis. These quantitative graph measures efficiently capture the cellular network organization, providing insights into their evolution, function, stability, and dynamic responses *(18)*. It has been demonstrated that protein–protein interaction, gene regulation, and metabolic networks exhibit scale-free topologies, characterized by a power-law degree distribution that decays slower than exponentially. A power-law network topology is frequently observed in different "real-world" networks and it can be generated by evolutionary models, where new nodes attach preferentially to sites that are already highly connected. Several modifications of this generic model, including iterative network duplication and integration to its original core, lead to hierarchical network topologies, which are characterized by nonconstant clustering coefficient distribution *(18, 19)*.

In practice, the architecture of large-scale, molecular networks is determined with sampling methods, resulting in subnets of the true network, and only these partial networks can be applied to characterize the topology of the underlying, hidden network *(20)*. It has been proved that it is possible to extrapolate from subnets to the properties of the whole network only if the degree distributions of the whole network and randomly sampled subnets have the same probability distribution *(21)*. While this is the case in specific classes of network graph models, including classical Erdös–Rényi and exponential random graphs, the condition is not satisfied for scale-free degree distributions. Accordingly, recent studies in interactome networks have revealed that the commonly accepted scale-free model for PPI networks may fail to fit the data *(22)*. Moreover, limited sampling alone may as well give rise to apparent scale-free topologies, irrespective of the original network topology *(23)*. These results suggest that interpretation of the global properties of the complete network structure based on the current – still limited – accuracy and coverage of the observed networks should be made with caution.

3.2. Subgraphs, Pathways, and Functional Modules

Opposed to the study of global characteristics of large-scale networks, local interconnectivity and more detailed measures were used to characterize local features of the molecular networks. Such graph methods can facilitate addressing fundamental biological concepts, such as essentiality and pathways, especially when additional biological information (i.e., *time series*) is incorporated into the analysis in addition to the primary data. Once the network of interest has been represented as a graph, the usual graph-driven analysis strategy is as following:

(1) Compute the local network properties (such as the number and complexity of given subgraphs, the shortest path length, or *centrality*) applying suitable graph algorithms.

(2) Evaluate the sensitivity and specificity of the model predictions using curated databases of known positive examples or random models of synthetic negative examples, respectively.

Centrality is a local quantitative measure of the position of a node relative to the other nodes, and can be used to estimate its relative importance or role in global network organization. Different flavors of centrality are based on the node's connectivity (degree centrality), its shortest paths to other nodes (closeness centrality), or the number of shortest paths going through the node (betweenness centrality). Estrada and colleagues demonstrated that centrality measures based on graph spectral properties can distinguish essential proteins in the protein–protein interaction network of *S. cerevisiae* (essential genes are those on which the cell depends for viability) *(24)*. In particular, the best performance in identifying essential proteins was obtained with a novel measure introduced to account for the participation of a given node in all subgraphs of the network (subgraph centrality), that gives more weight to smaller subgraphs. It was proposed that ranking proteins according to their centrality measures could offer a means to selecting possible targets for drug discovery *(24)*. A similar approach to characterize the importance of individual nodes, based on trees of shortest paths and concepts of "bottleneck" nodes, demonstrated that 70% of the top ten most frequent "bottleneck" proteins were inviable and structural proteins that do not participate in cellular signaling *(22)*. With degree centrality analyses in the metabolic networks of *E. coli*, *S. cerevisiae*, and *S. aureus*, it was also demonstrated that most reactions identified as essential turned out to be those involving the production or consumption of low-degree metabolites *(25)*.

In the theory of directed graphs, a path is a chain of distinct nodes, connected by directed edges, without branches or cycles. Such pathways in cellular network graphs can represent, for instance, a transformation path from a nutrient to an end-product in a metabolic network, or a chain of posttranslational modifications from the sensing of a signal to its intended target in a signal transduction network *(18)*. Pathway redundancy (the presence of multiple paths between the same pair of nodes) is an important local property that is thought to be one reason for the robustness of many cellular networks. Betweenness centrality can be used to measure the effect of node perturbations on pathway redundancy, whereas path lengths characterize the response times under perturbations. With a shortest paths and centrality-based predictions in the yeast protein–protein interaction and metabolic networks, respectively, the existence of alternate paths that bypass viable proteins can be demonstrated, whereas lethality corresponds to the lack of alternative pathways in the perturbed network *(22, 26)*. Besides the various commercial software packages for pathway analysis, there exist also freely available tools for some specific graph queries, such as finding shortest paths between two specified seed nodes on degree-weighted metabolic networks *(27)*, or searching for linear paths that are similar to query pathways in terms of the their composition and interaction patterns on a given protein–protein interaction network *(28)*.

The relatively high degree of noise inherent in the interactions data in high-throughput experiments makes pathway modeling very challenging. Integration of prior biological knowledge, such as Gene Ontology (GO), can be used to make the process of inferring models more robust by providing complementary information on protein function. GO terms and their relationships are encoded in the form of directed acyclic graphs (DAGs).

Guo et al. *(29)* recently assessed the capability of both GO-graph structure-based and information content-based similarity methods to evaluate protein–protein interactions involved in human regulatory pathways. They also showed how the functional similarity of proteins within known pathways decays rapidly as their path length increases. While most of the analysis methods designed for PPI networks consider unweighted graphs, where each pairwise interaction is considered equally important, Scott et al. *(30)* recently presented linear-time algorithms for finding paths and more general graph structures such as trees that can also consider different reliability scores for PPIs. By exploiting a powerful randomized graph algorithm, called color coding, they efficiently recovered several known yeast signaling pathways such as MAPK and showed that in general the pathways they detected score higher than those found in randomized networks. In addition to known pathways, they also predicted (by unsupervised learning) novel putative pathways in the yeast PPI network that are functionally enriched (i.e., share significant number of common GO annotations) *(30)*.

The decomposition of large networks into distinct components, or modules, has come to be regarded as a major approach to deal with the complexity of large cellular networks *(31)*. In cellular networks, a module refers to a group of physically or functionally connected biomolecules (nodes in graphs) that work together to achieve the desired cellular function *(19)*. To investigate the modularity of interaction networks, tools and measures have been developed that not only can identify whether a given network is modular or not, but can also detect the modules and their relationships in the network. By subsequently contrasting the found interaction patterns with other large-scale functional genomics data, it is possible to generate concrete hypotheses for the underlying mechanisms governing e.g., the signaling and regulatory pathways in a systematic and integrative fashion. For instance, interaction data together with mRNA expression data can be used to identify active subgraphs, that is, connected regions of the network that show significant changes in expression over particular subsets of experimental conditions *(32)*.

3.3. Motifs and Clusters

Motifs are subgraphs of complex networks that occur significantly more frequently in the given network than expected by chance alone *(33)*. Consequently, the basic steps of motif analyses are (1) estimating the frequencies of each subgraph in the observed network, (2) grouping them into subgraph classes consisting of isomorphic subgraphs (topologically equivalent motifs), (3) and determining which subgraph classes are displayed at much higher frequencies than in their random counterparts (under a specified random graph model).

While analytical calculations from random models can assist in the last step, exhaustive enumeration of all subgraphs with a given number of nodes in the observed network is impossible in practice. Kashtan et al. *(34)* therefore developed a probabilistic algorithm that allows estimation of subgraph densities, and thereby detection of network motifs, at a time complexity that is asymptotically independent of the network size. The algorithm is based on a subgraph importance sampling strategy, instead of standard Monte Carlo sampling. They noticed that network motifs could be detected already with a small number of samples in a wide variety of biological networks, such as the transcriptional regulatory network of *E. coli (34)*. Recently, efficient alternatives together with graphical user interfaces have also been implemented to facilitate fast network motif detection and visualization in large network graphs *(35, 36)*.

Many of the methodologies recently introduced in network analysis are inspired by established approaches from sequence analysis. The concepts utilized in both fields include approximate similarity, motifs, and alignments. As network motifs represent a higher-order biological structure than protein sequences, graph-based methods can be used to improve the homology detection of standard sequence-based algorithms, such as PSI-BLAST, by exploiting relationships between proteins and their sequence motif-based features in a bipartite graph representing protein-motif network *(37)*. The definition of network motifs can be enriched by concepts from probability theory. The motivation is that, if the network evolution involves elements of randomness and currently available interaction data is imperfect, then functionally related subgraphs do not need to be exactly identical. Accordingly, Berg and Lässig devised a local graph alignment algorithm, which is conceptually similar to sequence alignment methodologies *(38)*. The algorithm is based on a scoring function measuring the statistical significance for families of mutually similar, but not necessarily identical, subgraphs. They applied the algorithm to the gene regulatory network of *E. coli (38)*.

An alternative approach to the identification of functional modules in complex networks is discovering similarly or densely connected subgraphs of nodes (clusters), which are potentially involved in common cellular functions or protein complexes. As in expression clustering, the application of graph clustering is based on the assumption that a group of functionally related nodes are likely to highly interact with each other while being more separate from the rest of the network. The challenges of clustering network graphs are similar to those in the cluster analysis of gene expression data *(39)*. In particular, the results of most methods are highly sensitive to their parameters and to data quality, and the predicted modules need not be exactly similar, especially when the boundaries and connections between the modules are not clear-cut. This seems to be the case at least in the yeast PPI network *(17)*. Moreover, it should be noted that modules are generally not isolated components of the networks, but they share nodes, links, and even functions with other modules as well *(19)*. Such hierarchical organization of modules into smaller, perhaps overlapping and functionally more coherent modules should be considered when designing network clustering algorithms. The functional homogeneity of the nodes in a cluster with known annotations can be assessed against the cumulative hypergeometric distribution that represents the null model of random function label assignments *(22)*.

3.3.1. Highly Connected Clusters

Most algorithms for determining highly connected clusters in PPI networks yield disjoint modules *(40)*. For instance, King et al. partitioned the nodes of a given graph into distinct clusters, depending on their neighboring interactions, with a cost-based local search algorithm that resembles the tabu-search heuristic (i.e., it updates a list of already explored clusters that are forbidden in later iteration steps) *(41)*. Clusters with either low functional homogeneity, cluster size, or edge density were filtered out. After optimizing the filtering cut-off values according to the cluster properties of known yeast protein complexes from MIPS database, their methods could accurately detect known and predict new protein complexes *(41)*. Other local properties such as centrality measures can be used for clustering purposes as well. An algorithm by Dunn et al., for example, divides the network into clusters by removing the edges with the highest betweenness centralities, then recalculating the betweenness and repeating until a fixed number of edges have been removed *(42)*. They applied the clustering method to a set of human and yeast PPIs, and found that the protein clusters with significant enrichment for GO functional annotations included groups of proteins known to cooperate in cell metabolism *(42)*.

3.3.2. Overlapping Clusters

Corresponding to the fact that proteins frequently have multiple functions, some clustering approaches, such as the local search strategy by Farutin et al. also allow overlapping clusters *(43)*. Like in motif-analysis, the score for an individual cluster in the PPI network graph is assessed against a null model of random graph that preserves the expected node degrees. They also derived analytical expressions that allow for efficient statistical testing *(43)*. It was observed that many of the clusters on the human PPI network are enriched for groups of proteins without clear orthologs in lower organisms, suggesting functionally coherent modules *(43)*. Pereira-Leal et al. used the line graph of the network graph (where nodes represent an interaction between two proteins and edges shared interactors between interactions) to produce an overlapping graph partitioning of the original PPI network of

S. cerevisiae (44). Recently, Adamcsek et al. provided a program for locating and visualizing overlapping, densely interconnected groups of nodes in a given undirected graph *(45)*. The program interprets as motifs all the k-clique percolation clusters in the network (all nodes that can be reached via chains of adjacent k-cliques). Larger values of k provide smaller groups resulting in higher edge densities. Edge weights can additionally be used to filter out low-confidence connections in the graphs *(45)*.

3.3.3. Distance-Based Clusters

Another approach to decompose biological networks into modules applies standard clustering algorithms on vectors of nodes' attributes, such as their shortest path distances to other nodes *(46)*. As the output then typically consists of groups of similarly linked nodes, the approach can be seen as complementary to the above clustering strategies that aim at detecting highly connected subgraphs. To discover hierarchical relationships between modules of different sizes in PPI graphs, Arnau et al. explored the use of hierarchical clustering of proteins in conjunction with the pairwise path distances between the nodes *(47)*. They considered the problem of lacking resolution caused by the "small world" property (relatively short – and frequently identical – path length between any two nodes) by defining a new similarity measure on the basis of the stability of node pair assignments among alternative clustering solutions from resampled node sets. As ties in such bootstrapped distances are rare, standard hierarchical clustering algorithms yield clusters with a higher resolution. The clusters obtained in yeast PPI data were validated using GO annotations and compared to those refined from gene expression microarray data *(47)*. A similar approach was also applied to decompose the metabolic network of *E. coli* into functional modules, based on the global connectivity structure of the corresponding reaction graph *(48)*.

4. Data Integration Methods

The goal of information integration in science is to combine information from a number of databases and datasets under one data management scheme such that the cumulative information provides greater insight into domain problems than is possible with separate information sources. As the high-throughput assays are inherently noisy and biased in their nature, and each single data source or type can describe only a limited scope of a system, it is evident that integrative analysis of data from such measurements will be essential in order to fully understand the system's behavior on a global scale [ref]. In many biological applications, it is beneficial to perform the network analysis in a truly integrated manner, simultaneously rather than sequentially, like when validating the results against external data sources.

In recent years there has been a substantial increase in the number and variety of molecular interaction studies (see Gupta and Ludäscher *(49)*). Deciphering genome-wide protein–protein interactions (Li et al. *(50)*), large-scale analysis and prediction of gene regulatory networks (Vert and Kanehisa *(51)*), and construction of metabolic pathways (Famili and Palsson *(52)*) have become common in the past several years. Alongside, systems biologists also are discovering genetic interaction networks where two nodes (genes) are connected by an edge only if a mutation in any one of the genes has no phenotypic effect but simultaneous mutation of both genes causes a phenotype. However, there is relatively little research (exceptions are Mendes et al. *(8)* and Krishnamurthy et al. *(9)*) in the development of an information management framework that adequately models the graph-structured nature of the data and the graph-oriented operations that different users perform today. In the absence of an efficient information management system that allows biologists to query discrete and large databases simultaneously, the full potential of functional genomics resources remains underutilized. The goal of this section is to describe information integration methods as well graph-theoretic methods integrating molecular interaction networks and ontologies, and to show how they can be used by biologists to address biologically relevant questions about patterns of protein–protein and protein–DNA interactions.

We illustrate the utility of the graph information integration with a simple example. Several techniques in bioinformatics information management that represent data as graphs make use of clique structures to solve problems. To a biologist, the determination of a clique in protein–protein interaction networks is important for biological hypothesis generation, for example, in the formulation of putative modules of interacting proteins as the signature of a putative biochemical machine (Rives and Galitski *(53)*). Consider that the graph in **Fig. 2a** comes from source S1 and that in **Fig. 2b** comes from source S2. Now, if a query seeks a cliquish region in the graph having node count greater than 4, each source would individually return an empty result.

A number of other groups have developed data models for aspects of molecular interaction (often called "pathways"). Relatively fewer groups have published their schemas, and articulated their data modeling or schema design principles. In this short paper, we select a few example systems. The PathDB (Mendes et al. *(8)*) from the National Center of Genome Resources (www.ncgr.org/software/pathdb) is a relational database developed for metabolic networks. In PathDB, the central element is a biochem, which is classified as one of RNA, DNA, Compound, Protein,

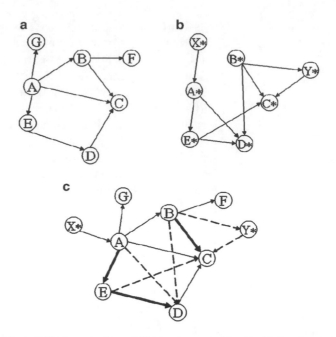

Fig. 2. The individual graphs (**a**) and (**b**) have been combined in (**c**). A node in graph (**a**) corresponds to *-marked node in graph (**b**). The *thin solid edges* come only from (**a**), the *dashed edges* come only from (**b**), while the *thick edges* are present in both.

or Subunit, and in turn used to build other biochem objects. The transition is modeled by the explicit representation of a biochemical reaction whose substrates, products, and mediator with its kinetic properties are recorded. The PathDB system does not provide a query language, but instead provides a Query Tool that allows a user to specify a large but prespecified set of queries of varying complexity. Fukuda and Takagi (see Fukuda and Takagi *(54)*) consider a pathway as a first-order logical structure called the compound graph. A compound graph is created when a tree and a graph are constructed over the same set of nodes – the tree represents the hierarchy of complexes and their constituent objects, while the graph represents relations like binds to between objects. The edge labels of a graph can be organized in a hierarchy – phosphorylation, for instance, is a special case of protein modification. The authors show how Prolog can be used to perform recursive traversals over tree and graph edges and construct intentional (rule-defined) data objects. It is not clear how the system scales with large graphs. The Pathways database system described in Krishnamurthy et al. *(9)* models pathways as a directed hypergraph where nodes represent pathway elements like substrates and products of a reaction. Because of its graph-structured modeling, this system can more easily support queries where operations such as shortest path, unions and intersections of paths, and finding node-neighborhoods can be performed. The system provides the users with a query tool rather than a language, thus limiting the kinds of questions that can be asked of the database. Other groups include Ochs et al. *(10)*, who developed a metabolic map from a relational model of biochemical interactions, and Bhalla *(11)*, where a database of chemical reactions is mapped to a system of pathway graphs. However, none of these systems provide any support for information integration or ad hoc declarative queries.

Typically, the task of integrating a new data source to an existing integrated graph schema consists of three steps – defining a new, unpopulated data source in the integrator, mapping the just-imported schema to nodes, node attributes, and edges of the integrated graph, and expressing conflict resolution policies.

4.1. Source Definition

Currently, the external data source can be a relational database schema, a tree-structured XML document, an RDF-styled triplet that describes an edge set of a graph, or a DAG-structured OWL (http://www.w3.org/TR/owl-features) document. Typically, a new ontology or a node/attribute type hierarchy, such as the phenotype classification tree from MIPS as shown in **Fig. 3**, will be presented to the system using a tree (here as an OWL description) data; on the other hand, a collection of node/edge instances and node properties are presented as relational data. To import this data one should first define a new data source.

where the newly imported data is nicknamed yeast phenotype. XML-RDF-OWL is a format known to the system. For a relational data source, we would declare the format as SQL. With the data source defined, now we specify a PathSys schema element for the new source.

CREATE TREE phenotype tree (version STRING VALUE '2.3',...)

SOURCE yeast phenotype;

Note that in PathSys, a tree is a special case of graph, that is internally used for query evaluation. In the next section, we describe how the newly imported data is mapped to the extant global schema of the system.

4.2. Schema Mapping

The task of schema mapping is to specify how an element of the imported source should be interpreted as an element of the internal schema of PathSys. We illustrate the problem by first considering the tree-structured source, and then a more elaborate relational example. In the first case, we would like the OWL schema to populate the node type hierarchy in **Fig. 3**. The mapping declarations are:

IMPORT NODE TYPE FROM yeast phonotype (Class as name,)

GRAPH phenotype tree

CREATE DATA SOURCE yeast phenotype (fullname 'Yeast Phenotype Classification',
reference 'localhost://phenotype.owl', description . . .)

format XML-RDF-OWL;

<owl:Class rdf:ID="starvation_sensitivity">

<rdfs:subClassOf>

<owl:Class rdf:ID="stress_response_defects"/>

</rdfs:subClassOf>

</owl:Class>

<owl:Class rdf:ID="phenotypes"/>

<owl:Class rdf:ID="G2_M_arrest">

<rdfs:subClassOf>

<owl:Class rdf:ID="cell_cycle_defects"/>

</rdfs:subClassOf>

</owl:Class>

<owl:Class rdf:about="#cell_cycle_defects">

<rdfs:subClassOf rdf:resource="#phenotypes"/>

</owl:Class>

<owl:Class rdf:ID="G1_arrest">

<rdfs:subClassOf rdf:resource="#cell_cycle_defects"/>

</owl:Class>

<owl:Class rdf:about="#stress_response_defects">

<rdfs:subClassOf rdf:resource="#phenotypes"/>

</owl:Class>

<owl:Class rdf:ID="heat_shock_sensitivity">

<rdfs:subClassOf rdf:resource="#stress_response_defects"/>

</owl:Class>

Fig. 3. A small fragment of the yeast phenotype classification from MIPS.

IMPORT RELATIONSHIP FROM yeast phenotype (subClassOf as child of)

GRAPH phenotype tree

The second example of source integration imports a relational schema (a fragment of the MIPS database) into the graph elements of the internal model. **Figure 4** shows the relational schema. In the MIPS complex relation, the attribute complex references the CId of the MIPS complex category relation. Notice that gene name is explicitly mapped to the preexisting attribute name. The expression (orf1, gene1) AS SOURCE REFERENCE states that for the edge the source node uses the attribute pair {orf1, gene1} as a foreign key. Also notice how the protein complexes and their members are defined explicitly as hypernodes and hypernode members.

4.2.1. Conflict Resolution

An interesting challenge encountered during the information integration process arises due to data conflicts. In our experience, most conflicts arise due to name or ORF mismatches. Two sources

Source Relations

MIPS genes (orf, gene name, coordinates, classification, description);
MIPS interaction (id, orf1, gene1, relationship, orf2, gene2, description, reference, evidence)
FOREIGN KEY (orf1, gene1) REFERENCES MIPS genes (orf, gene name),
FOREIGN KEY (orf2, gene2) REFERENCES MIPS genes (orf, gene name)
);
MIPS gene alias (orf, gene, alias); MIPS complex
category (cid, name);
MIPS complex (entry, ref, evidence, complex);

Mapping Directives

IMPORT NODE FROM MIPS genes (
orf AS ATTRIBUTE,
gene name AS ATTRIBUTE name,
coordinates AS ATTRIBUTE,
classification AS ATTRIBUTE,
description AS ATTRIBUTE
) TYPE gene;

IMPORT EDGE FROM MIPS interaction (
(orf1, gene1) AS SOURCE REFERENCE,
(orf2, gene2) AS TARGET REFERENCE, relationship AS ATTRIBUTE,
description AS ATTRIBUTE,
reference AS ATTRIBUTE,
evidence AS ATTRIBUTE
) DIRECTION undirected GRAPH mips graph;

IMPORT NODE ATTRIBUTE FROM MIPS gene alias (
(orf, gene) AS NODE REFERENCE,
alias AS ATTRIBUTE synonym
) GRAPH mips graph;

IMPORT HYPERNODE FROM MIPS complex category (
cid AS ATTRIBUTE, name AS ATTRIBUTE
) GRAPH mips graph PRIMARY IDENTIFIER (cid) SECONDARY IDENTIFIER (name);

IMPORT HYPERNODE MEMBER FROM MIPS complex (
complex AS NODE REFERENCE,
entry AS MEMBER NODE REFERENCE, ref AS MEMBER ATTRIBUTE,
evidence AS MEMBER ATTRIBUTE
) GRAPH mips graph;

Fig. 4. A small fragment of the MIPS relational data and its mapping.

may present genes having two different names but the same ORF. In cases like this, we have an automated reconciliation procedure – we assign multiple names as synonyms to the same ORF. More interestingly, consider three sources all seemingly referring to an item called "IME1." Let us say, all three records have an attribute called gene name, with values "IME1," "IME1-TAP (342–531)," and "IME1(modified:Thr:210)," respectively. The problem is that the first record refers to the gene IME1, the second to a fragment of the gene IME1 with the additional information that IME1 was detected by a method called TAP, and the third record refers to the protein encoded by IME1 (Ime1p) with the qualifier that the amino acid called Thr at the 210th position was modified, making it a variant protein. Thus, the records refer to objects that are not really equal, but must be correlated by creating additional constraints. Currently, reconciliation problems such as these are detected by a set of conflict detection rules (such as finding terms with a common prefix or suffix), but resolved by expert user intervention.

Once the new graph is integrated, the system computes all graph indexes for the new incoming graph and updates indices for the whole integrated graph.

4.3. Querying Graphs

BioNetSQL, the query language for interaction networks in the PathSys system, has the flavor of SQL that can be queried on bags of nodes, edges, and their attributes, but additionally allows the returned values to be bags of paths, trees, and graphs. Further, the language allows path, tree, and graph operations. A complete description of the language and the query evaluation process is beyond the scope of this paper. We will present features of the language through examples as it is used for scientific queries.

First, we pose a relational-style query – "Find colocalized proteins grouped by location," which shows more SQL-like features of the language.

SELECT p.location, set(p)

FROM yeastGraphDB G(N, E) WHERE p:N and p.type = 'protein'

GROUP BY p.location;

Here, the from clause refers to a graph. Further, thanks to the grouping condition, the output is a nested relation instead of a graph, where due to the inner structuring element set, this query produces a set of tuples (gene-pairs) for every binding of location.

In the next query, we use graph operations in the body of the query, and the return data type is a graph. "Find networks of colocalized proteins that are parts of some protein complex and are connected by either a 2-hybrid (y2h) edge or a comimmunoprecipitation (coIP) edge." SELECT graph(N2(n.name, n.source), E2(e.label, e.source))

FROM yeastGraphDB G1(N, E)

WHERE n:N and c:N and e:E and n.type << 'protein'and c.type = 'protein complex'

and (e.label = 'y2h' or e.label = 'coIP')

and pathExpr(G1, c//[member of]n) = true

The query declares a variable c whose type is protein complex. The query returns a graph, whose nodes n should be tuples with the attributes name and source (i.e., data source), and whose edges e should have a label and a source from which that edge was known. Recall that the system will convert this to a query on a connector node. The << operation specifies that the type of the node should be "under" protein in the node type hierarchy N. The last line should be read as "n has an edge whose label has the value member, and this edge points to c," where c is declared before. Note that we did not mention the relationship between nodes n and edges e, namely, an instance of the returned edge set e connects instances of the returned node set n. This constraint, expressed as n.edge = e, is implied by the construct of line 2, where n and e are constrained to be parts of the same graph.

A common operation performed with a molecular interaction graph is to "color" the nodes by Gene Ontology labels. Since the Gene Ontology has three main segments and ten levels, it is customary to specify the level by which the nodes are to be colored. To PathSys, this really means that we need to join the data graph with the Gene Ontology and add to each node, a new attribute corresponding to the term of the Ontology at the chosen level. Recall that the Gene Ontology is really a directed acyclic graph. In PathSys, a graph can be declared to be a DAG, and this allows the user to perform some DAG-specific operations on it. We have modeled Gene Ontology to be a single rooted DAG, having the root at "goRoot." In the following, we illustrate the query "Find the 2-neighborhood of the gene IME1, without considering any node or edge from the preBIND or BIND databases, and color the nodes with Level 5 nodes from the Molecular Function segment of Gene Ontology."

WITH ime1 nn AS (

WITH G2 AS

(SELECT graph(n, e) FROM yeastGraphDB G1(N, E) WHERE n:N and e:E and n.source NOT IN ('BIND-DB', 'PreBIND-DB') and (e.sourceNode = n or e.targetNode = n)

)

SELECT k-neighborhood(G2, n1, 4)

FROM G2(N2, E2)

WHERE n1:N2 and n1.type = 'gene' and n1.name = 'IME1')

SELECT graph(N3(n3.name, goTerms.name as go id), E3)

FROM ime1 nn(N3,E3), GeneOntology GO(N4, E4)

WHERE n3:N3 and goTerms:N4

and goTerms IN extractNnodes(GO.goRoot/'Molecular Function', 5)

and pathExpr(GO, goTerm//n3.name) = true

This query illustrates a number of graph functions supported by the system. The function k-neighborhood(const node, k) creates a graph – note that the value of k is 4 to accommodate the connector nodes. The nodes for any graph create a set of nodes belonging to the graph. The function extractNodes (DAGRoot, Level) takes the root of a DAG and a level number as input and returns a set of nodes from the DAG at the specified level. The pathexpr function takes a path expression as input and returns a set of bindings for the query variables (here GOTerm and n.name). Structurally, the query shows how ime1 nn is created as an intermediate graph variable, which is θ-joined with a path from the Gene Ontology graph.

Finally, we present an example from systems biological analysis that uses the graph-theoretic attributes. In this example, we take two subnetworks b1 and b2 produced by two subqueries, each using an aggregate graph function. For each network, the query computes the distribution of the betweenness centrality of the nodes of the respective graphs, and then uses the F-test to compare them.

WITH b1 AS (

SELECT distribution(betweenness centrality(*,0.05))

FROM yeastGraphDB G1(N, E)

WHERE n:N and e:E and n.source IN ('Gavin-DB','Ito-DB','Tong-DB')

and n.degree() >),

b2 AS (…)

SELECT F-test(b1, b2)

FROM b1, b2, stat-source;

Due to a heavy use of statistical operations, a number of statistical operations have been packaged in a source called stat-source. The function between-centrality produces a bag of values corresponding to the betweenness centrality of all nodes satisfying the remaining constraints. The function distribution takes a set of values and a bucket size and outputs a histogram, which is known to the system as a basic statistical data type defined as a table of 2-tuples {category, count} – here the category comes from the number of distinct values of the centrality measure with a bucket size of 0.05.

4.4. Software Tools for Graph-Based Network Analysis

The availability of genome-scale datasets has increased the need for software tools that can integrate, construct, analyze, and visualize the high-dimensional data effectively. Several such software packages available for these challenging tasks along with their specific functionalities were recently listed by Joyce and Palsson *(55)*. Publicly available software systems that use graph-based data integration frameworks include e.g., Cytoscape together with its recent plug-ins *(56–58)*, megNet *(59)*, BioPIXIE *(60)*, Pointillist *(61, 62)*, PIANA *(63)*, BiologicalNetworks and PathSys *(12, 13)*. An important component of such systems is the possibility to visualize the graphs under analysis. This can be regarded as a fundamental tool in explorative network analysis; even if one wants to address only a very specific question within the given network graph, it may be helpful to visualize the result to discern possible flaws or follow-up questions. Recently introduced graph drawing tools include e.g., WebInterViewer *(64)* and CADLIVE *(65)*. Also, the Bioconductor project incorporates open-source tools to support computational analysis of graphical data structures (http://www.bioconductor.org/). The available packages implement not only algorithms for efficient graph visualization (AT&T Graphviz), but also the C++ Boost Graph Library for basic graph algorithms (RBGL package). At present, procedures that can be interfaced in the R environment include minimum spanning tree construction, shortest path finding, depth-first search, topological sorting, edge-connectivity measurement, and connected component decomposition *(66)*. The number of packages available within Bioconductor grows rapidly as many authors make their R source codes freely available for academic use *(67–69)*. Besides advanced graph algorithms, future software tools should meet the challenges of automated construction and simultaneous visualization of multiple pathways, as well as relating them and their interconnections to the underlying biological significance.

5. Ontologies

The first and most natural reaction by people to overwhelming amounts of raw data being produced by new information gathering techniques is to attempt to categorize and classify. Classification (frequently organized in the form of a hierarchy) is one way in which people organize a domain in order to understand it more easily. Biology was the first discipline to engage in systematic, large-scale classification because of the enormous complexity of its domain. The eXtensible Markup Language (XML) is a powerful and flexible mechanism that is usually used to represent bioinformatic data and facilitates communication. What makes XML powerful is the ability to organize data hierarchically. For example, organization of information about biopolymers starting at the organism level and successively elaborating up to individual DNA bases could be represented as follows: the "*organism*" element contains "*chromosome*" elements, which, in turn, contain "*locus*" elements, which contain "*genes*," which contain the "*DNA sequence," "domains," "exons," "introns,"* and so on. Along the way, the elements also contain references to database entries that furnish the source material for the genomic information.

Each kind of element in an XML document represents a concept. Concepts are the means by which people understand the world around them. Individuals must have a shared conceptual framework in order to communicate, but communication requires more than just a shared conceptualization; it is also necessary for the concepts to have names, and these names must be known to the two individuals who are communicating.

Biochemistry has a rich set of concepts ranging from very generic notions such as "*chemical*" to exquisitely precise notions such as "tumor necrosis factor alpha-induced protein 3." Concepts are typically organized into hierarchies to capture at least some of the relationships between them. XML document hierarchies are a means by which one can represent such hierarchical organizations of knowledge. **Figure 5** illustrates a hierarchy of chemicals taken from EcoCyc (EcoCyc 2003). For example, protein is more specific than chemical, and enzyme is more specific than protein. Classifications that organize concepts according to whether concepts are more general or more specific are called taxonomies by analogy with biological classifications into species, genera, families, and so on.

Hierarchies are traditionally obtained by starting with a single all-inclusive class, such as "BioEntity," and then subdividing into more specific subclasses based on one or more common characteristics shared by the members of a subclass. These subclasses are, in turn, subdivided into still more specialized classes, and so on, until the most specific subclasses are identified. An alternative

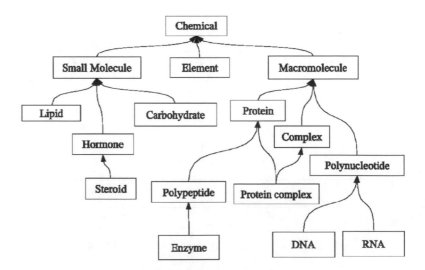

Fig. 5. Chemical hierarchy (EcoCyc 2003).

to the top-down technique is to start with the most specific classes. Collections of the classes that have features in common are grouped together to form larger, more general, classes. This is continued until one collects all of the classes together into a single, most general, class. This approach is called "bottom-up" classification. This is the approach that has been used in the classification of genes, called GeneOntology. Whether one uses a top-down or bottom-up technique, it is always presumed that one can define every class using shared common characteristics of the members.

Genes are classified both by function and by sequence. The two approaches interact with one another in complex ways, and the classification is continually changing as more is learned about gene function. **Figure 6** shows some examples of the classification of genes into families and superfamilies. The superfamily is used to describe a group of gene families whose members have a common evolutionary origin but differ with respect to other features between families. A gene family is a group of related genes encoding proteins differing at fewer than half their amino acid positions. Within each family there is a structure that indicates how closely related the genes are to one another. The relationships among the various concepts are complex, including evolution, duplication, and translocation.

The hierarchies shown in **Figs. 5** and **6** are very different from one another due to the variety of purposes represented in each case. The chemical hierarchy in **Fig. 5** is a specialization/

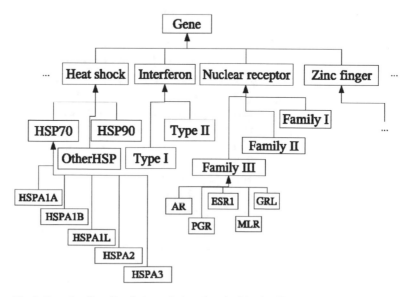

Fig. 6. Gene families. The first row below Gene in this classification consists of superfamilies. The row below that contains families. Below the families are some individual genes.

generalization hierarchy. The relationship here is called sub-class because mathematically it represents a subset relationship between the two concepts. The gene families and superfamilies in **Fig. 6** are also related by the subclass relationship, but the individual genes shown in the diagram are members (also called instances) of their respective families rather than being subsets. The relationships in this last diagram are neither subclass nor instance relationships but rather more complex relationships such as: evolves by mutation, duplicates, and translocates.

The organization of concepts into hierarchies can capture at least some of the relationships between them, and such a hierarchy can be represented using an XML document hierarchy. The relationship in an XML document between a parent element and one of its child elements is called containment. However, the actual relationship between the parent element and child element need not be a containment. For example, it is reasonable to regard a chromosome as containing a set of locus elements because a real chromosome actually does contain loci. Similarly, a gene really does contain exons, introns, and domains. However, the relationship between a gene (or interaction) and a reference where this gene (or interaction) was first described is not one of containment, but rather the referral or citation relationship. One of the disadvantages of XML is that containment is the only way to relate one element to another explicitly. The problem is that all the various kinds of hierarchy and various forms of relationship have to be represented using containment.- RDF, OWL The hierarchy in **Fig. 6** does not use any relationships that could reasonably be regarded as being containment. Yet, one must use the containment relationship to represent this hierarchy. The actual relationship is therefore necessarily implicit, and some auxiliary, informal technique must be used to elucidate which relationship is intended.

Unfortunately, this is not the only problem. One could not communicate very much if all one had were concepts and a single kind of relationship. Relating concepts to each other is fundamental. Linguistically, concepts are usually represented by nouns and relationships by verbs. Because relationships relate concepts to concepts, the linguistic notion of a simple sentence, with its subject, predicate, and object, represents a basic fact. The subject and object are the concepts and the predicate is the relationship that links them.

There are two ways to specify relationships in XML. The first technique is to add another "layer" between elements that specifies the relationship. This is called striping. A BioML document could be represented using striping. If one consistently inserts a relationship element between parent and child concept elements, then one can unambiguously distinguish the concept elements from the relationship elements. Striping was first introduced in the Resource Description Framework (RDF) (Lassila and Swick 1999).

Another way to specify a relationship is to use a reference. A reference is an attribute of an XML element that refers to some other data. The referenced data can be anything and anywhere, not just XML elements and not just in the same XML document. In general, a reference could be to anything that has a URI. This technique is much more flexible and powerful than striping.

Striping and references can be used in the same document. In RDF, the two techniques can be used interchangeably, and they have exactly the same meaning. A relationship specified with either striping or a reference forms a statement. Both striping and references help organize XML documents so that relationships are explicit. They contribute to the goal of ensuring that data are self-describing. References are commonly used in bioinformatics ontologies, but striping is seldom used outside of RDF ontologies.

One feature of RDF that makes it especially attractive is that its semantics have been formalized using mathematical logic. There are now a number of ontology languages that extend RDF and that also have formal semantics. The DARPA Agent Markup Language (DAML) is a DARPA project that produced the DAML + OIL language. This language has recently been superseded by the Web Ontology Language (OWL). OWL is a standard of the World Wide Web Consortium (W3C). The RDF and OWL standards are available on the W3C website (www.w3c.org).

In big biological ontologies the same term can be used in different ways. For example, "locus" is an attribute in the Bioinformatic Sequence Markup Language (BSML), but it is an element in BioML. The same term can have different meanings; this is especially true of commonly occurring terms such as "value" and "label." The same term might have the same use and meaning, but it may be constrained differently: for example, the "Gene" element occurs in several DTDs and has the same meaning, but the content and attributes that are allowed will vary. Namespaces were introduced to XML to allow one to use multiple DTDs or XML schemas without confusing the names of elements and attributes that have more than one meaning. A namespace is a URI that serves as means of distinguishing a set of terms. For example, reaction is used both in the Systems Biology Markup Language (SBML) (SBML 2003) and in CML. The SBML namespace is http://www.sbml.org/sbml/level2. The CML namespace dealing with chemical reaction terminology is http://www.xml-cml.org/schema/cml2/react

By using the namespaces one can ensure that any use of reaction is unambiguous. One can also declare a namespace to be the default namespace with unqualified element names belonging to the default namespace.

6. Notes

Here are examples of namespaces that are currently important for ontologies and frequently used by molecular and systems biologist:

bioml: Biopolymer Markup Language
http://xml.coverpages.org/bioMLDTD-19990324.txt
cellml: Cell Markup Language
http://www.cellml.org/cellml/1.0
cmeta: Cell Meta Language
http://www.cellml.org/metadata/1.0
cml: Chemical Markup Language
http://www.xml-cml.org/schema/cml2/core
dc: Dublin Core Elements
http://purl.org/dc/elements/1.1/
dcterms: Dublin Core Terms
http://purl.org/dc/terms/
go: Gene Ontology
http://ftp://ftp.geneontology.org/pub/go/xml/dtd/go.dtd
mathml: Mathematics Markup Language
http://www.w3.org/1998/Math/MathML
owl: Web Ontology Language
http://www.w3.org/2002/07/owl
rdf: RDF
http://www.w3.org/1999/02/22-rdf-syntax-ns
rdfs: RDF Schema
http://www.w3.org/2000/01/rdf-schema
sbml: Systems Biology Markup Language
http://www.sbml.org/sbml/level2
stm: Technical Markup Language
http://www.xml-cml.org/schema/stmml
xmlns: XML Namespaces
http://www.w3.org/XML/1998/namespace
xsd: XML Schema (original)
http://www.w3.org/2000/10/XMLSchema
xsd: XML Schema (proposed)
http://www.w3.org/2001/XMLSchema
xsi: XML Schema instances
http://www.w3.org/2001/XMLSchema-instance
xsl: XML Transform
http://www.w3.org/1999/XSL/Transform
xtm: Topic Maps
http://www.topicMaps.org/xtm/1.0/

7. Conclusion

Network graphs have the advantage that they are very simple to reason about, and correspond to the information that is globally available today. However, while binary relation information does represent a critical aspect of interaction networks, many biological processes appear to require more detailed models. Therefore, we expect that one of the main directions in the development of graph-based methods will be their extension to other types of large-scale data from existing and new experimental technologies. This may eventually prove mathematical models of large-scale datasets valuable in medical problems, such as identifying the key players and their relationships responsible for multifactorial behavior in human disease networks.

References

1. Small S, Blair A, and Levine M. (1992). Regulation of even-skipped stripe 2 in the Drosophila embryo. EMBO J 11(11):4047–4057.

2. Small S, Blair A, and Levine M. (1996). Regulation of two pair-rule stripes by a single enhancer in the Drosophila embryo. Dev Biol 175:314–324.

3. Yuh CH, Bolouri H, and Davidson EH. (1998). Genomic cis-regulatory logic: Experimental and computational analysis of a sea urchin gene. Science 279:1896–1902.

4. Madhani HD, and Fink GR. (1998). The riddle of MAP kinase signaling specificity. Trends Genet 14:4.

5. Davidson EH. (1986). Gene Activity in Early Development. Academic, Orlando, FL.

6. Kanehisa M, Goto S. (1999). KEGG: Kyoto encyclopedia of genes and genomes. Nucleic Acids Res 27:29–34. http://www.genome.ad.jp/kegg/.

7. Goto S, Bono H, Ogata H, Fujibuchi W, Nishioka T, Sato K, Kanehisha M. (1997). Organizing and computing metabolic pathway data in terms of binary relations. In Pacific Symposium on Biocomputing'97, pp. 175–186.

8. Blanchard JL, Bulmorc DL, Farmer AD, Gonzales M, Steadman PA, Waugh ME, Wlodek ST, and Mendes P. (2000). Pathdb: A second generation metabolic database. In Hofmeyr JH, Rohwer J, Snoep J. (eds.), Animating the cellular map, pp. 207–212. Stellenbosch University Press, Stellenbosch.

9. Krishnamurthy L, Nadeau J, Ozsoyoglu G, Ozsoyoglu M, Schaeffer G, Tasan M, Xu W. (2003). Pathways database system: An integrated system for biological pathways. Bioinformatics 19:930–937.

10. Ochs RA, Qureschi A, Sycz A, Vorbach J. (1996). A computerized metabolic map 2. relational structure, extended modeling and a graphical interface. J Chem Inf Comput Sci 36:594–601.

11. Bhalla US. (2002). The chemical organization of signaling interactions. Bioinformatics 18:855–863.

12. Baitaluk M, Qian X, Godbole S, Raval A, Ray A, and Gupta A. (2006). PathSys: Integrating molecular interaction graphs for systems biology. BMC Bioinformatics 7:55.

13. Baitaluk M, Sedova M, Ray A, and Gupta A. (2006). BiologicalNetworks: Visualization and analysis tool for systems biology. Nucleic Acids Res 34:W466–W471; doi:10.1093/nar/gkl308.

14. Lukashin AV, Lukashev ME, and Fuchs R. (2003). Topology of gene expression networks as revealed by data mining and modeling. Bioinformatics 19:1909–1916.

15. Klamt S, and Gilles ED. (2004). Minimal cut sets in biochemical reaction networks. Bioinformatics 20:226–234.

16. Wuchty S, Oltvai ZN, and Barabasi AL. (2003). Evolutionary conservation of motif constituents in the yeast protein interaction network. Nat Genet 35:176–179.

17. Han JD, Bertin N, Hao T, Goldberg DS, Berriz GF, et al. (2004). Evidence for dynamically organized modularity in the yeast protein-protein interaction network. Nature 430:88–93.

18. Albert R. (2005). Scale-free networks in cell biology. J Cell Sci 118:4947–4957.

19. Barabási AL, and Oltvai ZN. (2004). Network biology: Understanding the cell's functional organization. Nat Rev Genet 5:101–113.

20. Lappe M, and Holm L. (2004). Unraveling protein interaction networks with near-optimal efficiency. Nat Biotechnol 22:98–103.

21. Stumpf MP, Wiuf C, and May RM. (2005). Subnets of scale-free networks are not scale-free: Sampling properties of networks. Proc Natl Acad Sci USA 102:4221–4224.

22. Przulj N, Corneil DG, and Jurisica I. (2004). Modeling interactome: Scale-free or geometric? Bioinformatics 20:3508–3515.

23. Han J-DJ, Dupuy D, Bertin N, et al. (2005). Effect of sampling on topology predictions of protein-protein interaction networks. Nat Biotechnol 23:839–844.

24. Estrada E. (2006). Virtual identification of essential proteins within the protein interaction network of yeast. Proteomics 6:35–40.

25. Samal A, Singh S, Giri V, et al. (2006). Low degree metabolites explain essential reactions and enhance modularity in biological networks. BMC Bioinformatics 7:118.

26. Palumbo MC, Colosimo A, Giuliani A, et al. (2005). Functional essentiality from topology features in metabolic networks: A case study in yeast. FEBS Lett 579:4642–4646.

27. Croes D, Couche F, Wodak SJ, et al. (2005). Metabolic PathFinding: Inferring relevant pathways in biochemical networks. Nucleic Acids Res 33:W326–W330.

28. Shlomi T, Segal D, Ruppin E, et al. (2006). QPath: A method for querying pathways in a protein-protein interaction network. BMC Bioinformatics 7:199.

29. Guo X, Liu R, Shriver CD, et al. (2006). Assessing semantic similarity measures for the characterization of human regulatory pathways. Bioinformatics 22:967–973.

30. Scott J, Ideker T, Karp RM, et al. (2006). Efficient algorithms for detecting signaling pathways in protein interaction networks. J Comput Biol 13:133–144.

31. Hartwell LH, Hopfield JJ, Leibler S, et al. (1999). From molecular to modular cell biology. Nature 402(6761 Suppl):C47–C52.

32. Ideker T, Ozier O, Schwikowski B, et al. (2002). Discovering regulatory and signalling circuits in molecular interaction networks. Bioinformatics 18(Suppl 1):S233–S240.

33. Milo R, Shen-Orr S, Itzkovitz S, et al. (2002). Network motifs: Simple building blocks of complex networks. Science 298:824–827.

34. Kashtan N, Itzkovitz S, and Milo R. (2004). Efficient sampling algorithm for estimating subgraph concentrations and detecting network motifs. Bioinformatics 20(11):1746–1758.

35. Wernicke S, and Rasche F. (2006). FANMOD: A tool for fast network motif detection. Bioinformatics 22:1152–1153.

36. Schreiber F, and Schwobbermeyer H. (2005). MAVisto: A tool for the exploration of network motifs. Bioinformatics 21:3572–3574.

37. Kuang R, Weston J, Noble WS, and Leslie C. (2005). Motif-based protein ranking by network propagation. Bioinformatics 21:3711–3718.

38. Berg J, and Lässig M. (2004). Local graph alignment and motif search in biological networks. Proc Natl Acad Sci USA 101:14689–14694.

39. D'haeseleer P. (2005). How does gene expression clustering work? Nat Biotechnol 23:1499–1501.

40. Brun C, Herrmann C, and Guenoche A. (2004). Clustering proteins from interaction networks for the prediction of cellular functions. BMC Bioinformatics 5:95.

41. King AD, Przulj N, and Jurisica I. (2004). Protein complex prediction via cost-based clustering. Bioinformatics 20:3013–3020.

42. Dunn R, Dudbridge F, and Sanderson CM. (2005). The use of edge-betweenness clustering to investigate biological function in protein interaction networks. BMC Bioinformatics 6:39.

43. Farutin V, Robison K, Lightcap E, et al. (2006). Edge-count probabilities for the identification of local protein communities and their organization. Proteins 62:800–818.

44. Pereira-Leal JB, Enright AJ, and Ouzounis CA. (2004). Detection of functional modules from protein interaction networks. Proteins 54:49–57.

45. Adamcsek B, Palla G, Farkas IJ, et al. (2006). CFinder: Locating cliques and overlapping modules in biological networks. Bioinformatics 22(8):1021–1023.

46. Rives AW, and Galitski T. (2003). Modular organization of cellular networks. Proc Natl Acad Sci USA 100:1128–1133.

47. Arnau V, Mars S, and Marin I. (2005). Iterative cluster analysis of protein interaction data. Bioinformatics 21:364–378.

48. Ma HW, Zhao XM, Yuan YJ, et al. (2004). Decomposition of metabolic network into functional modules based on the global connectivity structure of reaction graph. Bioinformatics 20:1870–1876.

49. Gupta A, and Ludäscher B. (2003). The many faces of process interaction graphs: A data management perspective. OMICS 7:105–108.

50. Li S, Armstrong CM, Bertin N, Ge H, Milstein S, Boxem M, Vidalain PO, Han JD, Chesneau A, Hao T, Goldberg DS, Li N, Martinez M, Rual JF, Lamesch P, Xu L, Tewari M, Wong SL, Zhang LV, Berriz GF, Jacotot L, Vaglio P, Reboul J, Hirozane-Kishikawa T, Li Q, Gabel HW, Elewa A, Baumgartner B, Rose DJ, Yu

H, Bosak S, Sequerra R, Fraser A, Mango SE, Saxton WM, Strome S, Van Den Heuvel S, Piano F, Vandenhaute J, Sardet C, Gerstein M, Doucette-Stamm L, Gunsalus KC, Harper JW, Cusick ME, Roth FP, Hill DE, and Vidal M. (2004). A map of the interactome network of the metazoan *C. elegans*. Science 303:540–543.

51. Vert JP, and Kanehisa M. (2003). Extracting active pathways from gene expression data. Bioinformatics 19:238–244.

52. Famili I, and Palsson BO. (2003). Systemic metabolic reactions are obtained by singular value decomposition of genome-scale stoichiometric matrices. J Theor Biol 224:8796.

53. Rives AW, and Galitski T. (2003). Modular organization of cellular networks. Proc Natl Acad Sci USA 100:1128–1133.

54. Fukuda K, and Takagi T. (2001). Knowledge representation of signal transduction pathways. Bioinformatics 17:829–837.

55. Joyce AR, and Palsson BO. (2006). The model organism as a system: Integrating 'omics' data sets. Nat Rev Mol Cell Biol 7:198–210.

56. Shannon P, Markiel A, Ozier O, et al. (2003). Cytoscape: A software environment for integrated models of biomolecular interaction networks. Genome Res 13:2498–2504.

57. Reiss DJ, Avila-Campillo I, Thorsson V, et al. (2005). Tools enabling the elucidation of molecular pathways active in human disease: Application to hepatitis C virus infection. BMC Bioinformatics 6:154.

58. Albrecht M, Huthmacher C, Tosatto SC, et al. (2005). Decomposing protein networks into domain-domain interactions. Bioinformatics 21(Suppl 2):ii220–ii221.

59. Gopalacharyulu PV, Lindfors E, Bounsaythip C, et al. (2005). Data integration and visualization system for enabling conceptual biology. Bioinformatics 21(Suppl 1):i177 i185.

60. Myers CL, Robson D, Wible A, et al. (2005). Discovery of biological networks from diverse functional genomic data. Genome Biol 6:R114.

61. Hwang D, Rust AG, Ramsey S, et al. (2005). A data integration methodology for systems biology. Proc Natl Acad Sci USA 102:17296–17301.

62. Hwang D, Smith JJ, Leslie DM, et al. (2005). A data integration methodology for systems biology: Experimental verification. Proc Natl Acad Sci USA 102:17302–17307.

63. Aragues R, Jaeggi D, and Oliva B. (2006). PIANA: Protein interactions and network analysis. Bioinformatics 22:1015–1017.

64. Han K, Ju BH, and Jung H. (2004). WebInterViewer: Visualizing and analyzing molecular interaction networks. Nucleic Acids Res 32:W89–W95.

65. Li W, and Kurata H. (2005). A grid layout algorithm for automatic drawing of biochemical networks. Bioinformatics 21:2036–2042.

66. Carey VJ, Gentry J, Whalen E, et al. (2005). Network structures and algorithms in Bioconductor. Bioinformatics 21:135–136.

67. Scholtens D, Vidal M, and Gentleman R. (2005). Local modeling of global interactome networks. Bioinformatics 21:3548–3557.

68. Balasubramanian R, LaFramboise T, Scholtens D, et al. (2004). A graph-theoretic approach to testing associations between disparate sources of functional genomics data. Bioinformatics 20:3353–3362.

69. Zhu D, Hero AO, Cheng H, et al. (2005). Network constrained clustering for gene microarray data. Bioinformatics 21:4014–4020.

Chapter 5

Current Computational Methods for Prioritizing Candidate Regulatory Polymorphisms

Stephen Montgomery

Summary

Discovery of DNA sequence variants responsible for human phenotypic variation is key to advances in molecular diagnostics and medicines. Historically, variants that alter the protein-coding sequence of genes have been targeted when attempting to identify a trait's etiology; this is done because the rules governing these regions are generally well-understood and candidate variants can be easily selected. However, the effects of variants on gene regulation are increasingly regarded as being as important as protein-coding variation in uncovering the nature of phenotypic variation. I discuss resources and methodology that have recently been developed to computationally prioritize variants that may alter gene expression.

Key words: Regulatory polymorphisms, SNPs, Gene regulation, Transcription factor binding sites, Natural selection, Databases, Annotation

1. Introduction

Identification of the mechanisms by which genes are regulated in eukaryotes is one of the principal challenges of modern biology. The emergence of genome sequencing has facilitated the marked expansion of experimental and computational approaches designed to address this challenge. Integrating and assessing this information remains a major scientific endeavor that requires new and innovative application of technology. Furthermore, our limited understanding of the mechanisms of gene regulation in eukaryotes has undermined our ability to understand the role of genetics in gene regulation.

Vadim Astakhov (ed.), *Biomedical Informatics,* Methods in Molecular Biology, vol. 569
DOI 10.1007/978-1-59745-524-4_5, © Humana Press, a part of Springer Science+Business Media, LLC 2009

Regulatory variants are regarded as fundamental determinants of an organism's biological phenotype and fitness (for reviews, read *(1–5)*). Disease association studies have implicated them in the etiology of cancer *(6, 7)*, depression *(8)*, systemic lupus erythematosus *(9)*, perinatal HIV-1 transmission *(10)*, and response to type-1 interferons *(11)*. Whole genome association of gene expression in genotyped human populations and surveys of allelic-specific expression of transcripts have further supported their significance by reporting an abundance of polymorphisms associated with gene expression changes *(12, 13)*. However, experimental characterization of the mechanisms by which genes are expressed and the subsequent identification of the molecular cause and consequence of differential expression remains time-consuming and difficult. Computational approaches are uniquely suited to the problem of candidate regulatory polymorphism prioritization.

1.1. Patterns of Genetic Variation in Noncoding Regions

Genetic variation affecting gene regulation is manifested in *cis* through variation altering either transcription-factor binding or transcript viability or in *trans* through upstream mutation in transcription-mediating proteins (**Fig. 1**). Recently, the availability of single nucleotide polymorphism data derived from multiple populations and organisms has provided an essential resource for investigating the landscape of inheritance in the noncoding regions of genomes, key features of which have included defining the extent of linkage disequilibrium within a genome and identifying regions under selection.

Linkage disequilibrium in a genomic region is an important observation that defines the location within which a causal variant exists. It derives from the observation that one allele at one location can be informative for the presence of other alleles since linkage disequilibrium is the artifact of a new allele rising in frequency while bound to those preexisting alleles in which it originally occurred. The set of alleles that are co-inherited together is commonly defined as a "haplotype." Over time, subsequent additional mutation and recombination reduces the extent of correlation between variants creating smaller haplotypes. The purpose of association studies is to identify these haplotypes which contain both the causal variant and the associated variants linked to it; these studies have been limited in the past by the extent of variation that could be sampled and the extent of redundancy in sampling due to unknown linkage disequilibrium patterns. However, the recent availability of large-scale genotyping has made it practical to produce genome-wide maps of linkage disequilibrium. International projects like the HapMap have assayed over 3.1 million single nucleotide polymorphisms from four geographically distinct populations and have reported capturing at least 25–35% of common variation *(14)*. This data by design has not only supported the development of genome-wide

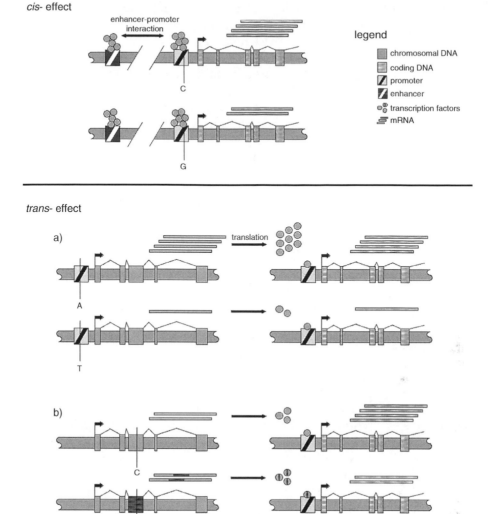

Fig. 1. *Cis-* and *trans-*regulatory interactions. *Top:* A C/G SNP in a transcription factor gene promoter alters the production rate of mRNA as a *cis-*effect. *Bottom:* An A/T SNP creating a *cis-*effect in a transcription factor gene alters in an allele-specific manner the production rate of mRNA in a downstream gene as a *trans-*effect. Additionally, a C/G coding SNP creates an altered transcription factor product that alters in an allele-specific manner the production rate of mRNA in a downstream gene as a *trans-*effect.

association studies by supplying testable target SNPs to test to ensure appropriate coverage, but also provided a resource for the evaluation of mutational forces such as recombination rates and signals of natural selection across the genome.

Studies of patterns of selection across the genome have recently taken advantage of haplotype data to identify those variants which have conferred a selective advantage in humans; it is thought that these variants are able to describe historical and ongoing human adaptation to their environment and, as such, likely carry functional variation *(15)*. Identifying these signals of selection has relied primarily on computing the rise in frequency of a derived allele on a time scale in contrast to that expected by random genetic drift such that further mutation and recombination has had insufficient time to break down the carrier haplotype. The Long Range Haplotype (LRH) test and integrated Haplotype Score (iHS) both test this feature by calculating the probability that two randomly chosen chromosomes carrying the target haplotype are identical by descent to a set genetic distance (typical where the probability reaches 0.05) against a controlled recombination rate background and in correlation to a rise in the derived allele's frequency *(16, 17)*. A more recent test, called the Cross-Population Extended Haplotype Homozygosity test, expands on both the LRH and iHS tests to identify whether a selected allele has approached fixation in one subpopulation but remains polymorphic in the human population as a whole *(18)*. The benefit of this is that population-specific alleles undergoing positive selection can be potentially determined. For instance, the authors report identifying population-specific selection in processes involving disease resistance and development.

1.2. The Role and Extent of Regulatory Polymorphism in Humans

It has been postulated that the differences between humans and chimpanzees are primarily due to regulatory variation since almost all key structural proteins remain virtually identical between the two species *(19)*. Furthermore, the recent sequencing of the human genome has identified fewer genes than were originally expected, suggesting a principal evolutionary role of alternative mechanisms, such as gene regulation and alternative splicing *(20, 21)*. The role of polymorphisms in such regions in determining the phenotypic diversity within a population has been of much interest. Quantification of the extent of regulatory variation has been predominantly explored using experimental assays designed to detect allele-specific gene expression.

Among the earliest techniques used to detect allele-specific changes in gene expression were reporter gene assays. Specific mutations are introduced in the promoter in question to assess their impact on expression. This technique has been recently used in three independent cell lines to estimate the impact of regulatory polymorphism on genes within a genome; it was found from a population of 170 genes that 35% contain regulatory polymorphisms *(22)*. The disadvantages of this technique are that it typically does not have the power to detect small differences in expression level, is laborious for high-throughput screening, and typically only one cell line or condition is assayed *(3)*. For verification purposes, many reporter gene analyses independently confirm allele-specific effects by assaying for differential protein binding using electrophoretic mobility shift assays.

Characterization of genes with allele-specific gene expression has also been focused on using techniques which identify genes with unbalanced expression of allelic transcripts in heterozygous samples *(2, 13)*. By identifying particular transcripts that are consistently expressed more frequently than others, it is assumed that each transcript is in linkage disequilibrium with an associated regulatory polymorphism. This approach has demonstrated its efficacy in identifying lowly-expressed transcripts in monogenic diseases such as Marfan syndrome *(23)*. In a series of high-throughput assays in normal tissues and conditions, it was predicted that between 25 and 50% of genes have allele-specific expression patterns using this approach *(1)*.

More recent techniques have taken advantage of advances in the scalability of expression technology to assess expression levels as a phenotype for linkage or association analyses *(10, 12, 24–26)*. These studies have detected what are termed expression quantitative trait loci (eQTLs) by identifying SNPs with significant linkage or association to changes in expression levels. These types of approaches, though, are not without their caveats as expression technology and cell culture introduce noticeable experimental variation when studies are compared to each other *(27)*.

1.3. Computation Methods for Prioritization of Regulatory Polymorphisms

Very few computational strategies have been developed to identify and prioritize candidate regulatory variants. SNP characterization tools, such as PupaSNP *(28)* and PolyMAPr *(29)*, identify putative regulatory SNPs by comparing their allele-specific predictions of transcription factor binding matrices stored in the TRANSFAC or JASPAR database *(30, 31)*. This approach has been used for genome-wide identification of polymorphisms that disrupt well-characterized consensus sequences; it demonstrated significant utility in locating regulatory variants within p53 response elements near the transcription start sites (TSS) of genes in the p53 response pathway as eight out of eight polymorphisms tested demonstrated function *(32)*. When coupled with phylogenetic footprinting between mouse and human, this technique has further demonstrated its utility on a set where seven out of ten SNPs that showed significant allele-specific differences in JASPAR database predictions also demonstrated electrophoretic mobility shift differences. However, only two of the seven had marked effect in reporter gene assays *(33)*. Furthermore, an analysis of 18 functional haplotypes and 30 nonfunctional haplotypes identified no enrichment in human–mouse conserved binding site predictions overlapping associated variants *(22)*. To facilitate further analysis using this technique, a database called PromoLign makes available SNPs and precomputed conserved binding sites between human and mouse for further analyses *(14)*. While neither study was statistically definitive, clearly, this suggests some effectiveness of approaches using database-driven regulatory prediction comparisons alone.

In another recently published study, the authors performed a statistical analysis of sequence composition from a collection of known regulatory mutations. The authors hypothesized that the composition around functional SNPs should be different than that around nonfunctional SNPs due to their selective roles in transcription factor binding (34). They observed that functional mutations of type C-to-T are slightly more associated with DNA regions with lower average sequence complexity with respect to symmetric elements and they attribute this to the known dyad symmetry of some transcription factor binding sites. In promoter regions, this technique was reported to have 70% specificity and 20% sensitivity. However, the ability to discriminate intragenic polymorphisms for functional importance was no better than random which was attributed to the absence of promoter-specific sequence composition differences. Of further consequence, their positive control set was of limited statistical power, as only 44 polymorphisms could be confidently characterized as functional. It is this paucity of known functional regulatory polymorphism and the likewise uncertainty of known nonfunctional SNPs which significantly challenges the development of robust discrimination techniques.

To address the lack of known regulatory polymorphisms, databases like MutDB (35), EnsEMBL (36), and the UCSC Genome Browser (37) show conservation profiles with polymorphism data and databases such as rSNP_Guide (38) and HGMD (39) catalogue regulatory mutations. rSNP_Guide further provides software for predicting the target transcription factors by coupling TFBS recognition scores from their database with user-supplied electrophoretic mobility shift assay data. However, the cumulative total of germline regulatory polymorphisms, which cause gene expression changes within the rSNP_Guide and HGMD databases, is approximately 60 and it is often difficult to discern what experimental conditions were originally used to predict them. Recently, community-based databases such as ORegAnno and PAZAR have been developed to provide standards for and encourage the annotation of regulatory elements and mutations (40, 41). These databases provide cell-line, expression, and experimental data and have since increased the number of annotated causal regulatory polymorphisms by threefold.

Using the regulatory polymorphisms annotated in ORegAnno we recently conducted a study of discriminating features of these regulatory SNPs by comparing them to colocalized SNPs of unknown function from dbSNP (42). This study was challenged by ascertainment bias due to the literature-derived nature of the tested known regulatory polymorphisms. However, it suggested several characteristic features of causal regulatory SNPs; they are close to the transcription start site, not within CpG islands, isolated from repetitive elements, possess higher MAF and higher derived allele frequency, and are within comparatively more divergent regions when comparing eukaryotic whole genome alignments. Furthermore, it highlighted that future studies will require polymorphisms that are nonfunctional across a broad range of cell types since recent analysis of allelic expression difference has demonstrated that the effects of rSNPs may be highly context-specific such that function in one cell line may not imply function in others *(37)*.

Population- or evolutionary-based identification of selection pressure on regulatory polymorphisms offers a complementary approach to these methodologies. It has been demonstrated that allele frequency shifts in human since primate divergence can detect functionally constrained regions *(43)*. Furthermore, when assayed in 114 human genes, allele frequencies below 6% have been observed to be enriched in functional polymorphisms suggesting that a simple filter of rare alleles enriches for functional regulatory SNPs *(14)*. Both of these studies suggest that frequency-based discrimination of functional regulatory polymorphisms should aid other computational approaches.

A purely computational approach to detecting allele-specific expression difference has been conducted through mining publicly available EST data *(10)*. SNPs in ESTs that were observed at nonequimolar ratios were assumed to be in linkage disequilibrium with a regulatory polymorphism. When tested, the authors were able to identify allele-specific expression changes in 36% of the genes, not overly different from experimental predictions *(1)*. But, of further significance in this study, the expression changes were identified to be in common SNPs, had been derived from multiple tissue sources, and showed greater consistency in allele-specific expression results.

A significant challenge to computational approaches aimed at detecting functional regulatory variation is a lack of understanding of the evolutionary history of regulatory regions. The majority of gene expression variation appears to follow a neutral model of evolution which suggests that most changes are due to stochastic processes *(5, 44)*. It has been further observed that 32–40% of the human functional sites are not functional in rodents meaning that regulatory evolution and selection may need to be analyzed through more evolutionarily-related species *(12)*. Specific observations in closely related species of purple sea urchin have identified considerable variation among known transcription factor binding sites compared to flanking sites, suggesting in some situations, sequence conservation is not always necessary for evolutionary retention of function *(45, 46)*.

2. Materials

Current computational methodology designed to prioritize candidate regulatory polymorphisms is based on associating regulatory features and predictions with information from observed mutations. Unlike in the coding regions of genomes, where a mutation's effect on codon structure is indicative of its viability, very few rules can be applied to preferentially predict the likelihood of a functional effect in noncoding regions. Furthermore, as our knowledge of the spatial and temporal activation of transcription factors and the genomic location of transcription factor binding is sparse, the relationship between sequence and function must be inferred. The following subsections describe the most common ways to predict the location of transcription factor binding sites and obtain polymorphism data and its associated frequency and derived allele status.

2.1. Gene Regulation Predictions and Annotation

2.1.1. Predicting Transcription Factor Binding Sites

Methodologies for predicting the locations of transcription factor binding sites can be broken down into five major classes, namely, those that: (1) use a signal-based approach, where a promoter or transcription factor binding site is determined from sequence composition; (2) use a database-driven approach, where new predictions are made from previously constructed promoters or transcription factor binding models; (3) use comparative genomics, where sequence conservation over multiple organisms implies functional constraint; (4) use function-based information (commonly from coexpression and gene ontology data); and (5) use architectural features of regulatory regions, such as combinatorial binding patterns or DNA curvature. These methodologies need not be mutually exclusive as comparative and coexpression-based approaches typically depend on de novo motif discovery algorithms or preexisting transcription factor binding models.

Of these methodologies, the most widely used way of identifying potential transcription factor binding sites is through matrix-based transcription factor binding site profiles. To facilitate this, transcription factor binding site databases, like TRANSFAC and JASPAR, have been developed to describe the diversity of sequences bound by a single transcription factor *(30, 31)*. Each database has benefited from in vitro binding assays, such as SELEX experiments (reviewed in *(47)*), to determine the sequence targets of transcription factors. The utility of this information is that a transcription factor's ability to bind specific sequences can then be modeled from the ensemble of possible binding sites and subsequently applied to novel sequences.

Transcription factor binding models within TRANSFAC and JASPAR are represented as either an IUPAC consensus sequence or a position-specific weight matrix. IUPAC consensus sequences were originally used to describe mutability between base positions of a transcription factor binding site by representing variant nucleotides with an enriched set of symbols. The disadvantage of this encoding is that the quantitative predisposition of individual bases to promote binding is essentially ignored; weight matrices were introduced to include this information *(48)*. A weight matrix is assembled by measuring the frequencies of individual nucleotides at each position in a binding site.

For working with position-specific weight matrices, the TFBS-Perl package provides a programming interface to databases like TRANSFAC and JASPAR which contain the matrix information, and allows the generation of representative sequence logos *(49)*. Alternately, sequence logos can also be constructed online using WebLogo *(50)*.

2.1.2. Gene Regulatory Annotation Resources

Despite the current lack of functional annotation regarding the regulatory architecture of mammalian genomes, several databases are available that aim to provide this information as it becomes available. The types of annotation that are recorded can be broken into three general classes: information regarding (1) the transcription factors and their interactions, (2) the locations of regulatory regions (promoters and enhancers), and (3) the locations of transcription factor binding sites. Many of the available resources are listed in **Table 1**.

Table 1
Gene regulation databases

The *Arabidopsis thaliana cis*-regulatory element database (AtcisDB) *(62)*	AtcisDB consists of transcription factor binding site information, promoter sequence, and related annotations for *Arabidopsis thaliana*. Within it, core promoters are predicted from full-length cDNAs. It contains 25,806 promoter sequences
Arabidopsis thaliana Promoter Binding Element Database (AtProbe) *(63)*	AtProbe consists of regulatory element information for *Arabidopsis thaliana* from Entrez, PlantCARE, PLACE, PubMed, and TRANSFAC. It contains 172 binding sites for 118 binding elements
Arabidopsis thaliana transcription factor database (AtTFDB) *(62)*	AtTFDB contains 1,690 *Arabidopsis* transcription factors and their sequences (protein and DNA) grouped into 50 (October 2005) families with information on available mutants in the corresponding genes
C. elegans promoter database (CEPDB) *(64)*	CEPDB contains promoters for 618 *C. elegans* genes as of June 2006
DBD: Transcription factor prediction database *(65)*	DBD contains transcription factor predictions for organisms based on homology through HMM modeling of domains. It contains predicted transcription factors for 150 completely sequenced genomes (37,736 transcription factors)

(continued)

Table 1
(continued)

Drosophila DNase1 Footprint Database *(40)*	A dataset based on a systematic literature curation and genome annotation of DNaseI footprints for *D. melanogaster*. It contains 1,367 binding sites for 87 transcription factors in 101 target genes from 201 primary references
The *Drosophila* transcription factor database (FlyTF) *(68)*	FlyTF is a database of fruitfly transcription factors. It contains 753 putative site-specific transcription factors of which 454 are well-supported
EdgeDB *(66)*	EdgeDB is a *C. elegans* differential gene expression database. It includes information regarding protein–protein and protein–DNA interactions including expression patterns conferred by regulatory elements
Eukaryotic Promoter Database (EPD) *(67)*	EPD is an annotated nonredundant collection of eukaryotic POL II promoters, for which the TSS has been determined experimentally
Hematopoiesis Promoter Database (HemoPDB) *(69)*	HemoPDB is composed of experimentally defined regulatory information, including TFs, *cis*-regulatory elements, their target gene promoters and corresponding annotations, with links to supporting published references with respect to hematopoiesis
JASPAR *(31)*	An alternative to TRANSFAC, this database is open access. JASPAR contains tightly controlled binding profiles with strict quality restrictions. Furthermore, it provides a programming API for ease of data access
The Liver Specific Gene Promoter Database (LSPD) *(70)*	LSPD contains liver-specific promoters and transcription factor binding sites. It contains 178 specific genes listed with 368 regulatory elements
Mammalian Promoter Database *(14)*	A collection of promoter databases for human, mouse, and rat
MPromDB *(14)*	MPromDB is a database for gene promoters with experimentally supported annotation of transcription start sites (TSS), *cis*-regulatory elements, CpG islands, and ChIP–chip experimental results
The Open Regulatory Annotation database (ORegAnno) *(40)*	ORegAnno is an open-access database of community-curated regulatory regions, transcription factors, and regulatory mutations
Orthologous Mammalian Gene Promoter database (OMGProm) *(71)*	OMGProm is a resource of mammalian gene promoters and their orthologs between humans and rodents
Osteo-Promoter Database (OPD) *(72)*	OPD is a database that contains analyses of promoters of genes in the osteogenic pathway
PAZAR *(41)*	PAZAR is an open-access, community-annotated database of transcription factor and regulatory sequence annotation
PLACE *(73)*	PLACE is a database of motifs found in plant *cis*-acting regulatory DNA elements as annotated from literature. It contains 469 entries

(continued)

Table 1
(continued)

PlantCARE *(74)*	PlantCARE is a database of plant *cis*-acting regulatory elements, enhancers, and repressors
PlantProm DB *(75)*	PlantProm DB is an annotated, nonredundant collection of proximal promoter sequences for RNA polymerase II with experimentally determined TSS from various plant species
Promoter Database of *Saccharomyces cerevisiae* (SCPD) *(76)*	SCPD is a yeast promoter database containing multiple promoters and transcription factor binding sites
Regulatory Element Database for *Drosophila melanogaster* (REDfly) *(40)*	REDfly is a curated collection of known *Drosophila* transcriptional *cis*-regulatory modules (CRMs). It contains 665 regulatory elements as of release 2.0
Riken Transcription Factor Database (TFdb) *(77)*	TFdb contains nonredundant transcription factors predicted for mouse
Transcriptional Regulatory Element Database (TRED) *(14)*	TRED is a mammalian regulatory element database. It contains genome-wide predictions of core promoters for human, mouse, and rat. It also contains expert-curated transcription factor binding sites for cell-cycle factors either computationally or experimentally determined
Transcriptional Regulatory Regions Database (TRRD) *(78)*	TRRD contains information on structural and functional organization of transcription regulatory regions of eukaryotic genes. It contains over 10,000 transcription factor binding sites and 3,490 regulatory regions curated from 7,609 references
TRANSFAC *(30)*	TRANSFAC is a widely used transcription factor site and matrix database. Within it, curators annotate transcription factor binding sites from literature to assemble representative consensus sequences and position weight matrices. However, binding profiles are of diverse quality and possess some redundancies

2.2. Single Nucleotide Polymorphism Resources

The central repository of polymorphism data is the dbSNP database at the National Center for Biotechnology Information in Washington, DC *(14)*. dbSNP curates polymorphism data through submission from researchers worldwide and assembles them into nonredundant entries. Several other databases have appeared with more specific mandates with respect to genetic mutation; some of these are listed in **Table 2**.

Among those databases not listed in **Table 2**, several databases have established themselves with a focus on polymorphisms relevant to a specific disease. Among these databases are the Breast Cancer Mutation database *(51)*, Cancer Genome Anatomy Project SNP index *(52)*, and the Cystic Fibrosis Mutation database *(53)*. The value of these databases is that typically rarer mutations identified from disease populations are differentiated from common mutations and individual variants can be tested for their role in disease.

Table 2
Gene mutation databases

The Single Nucleotide Polymorphism database (dbSNP) *(14)*	dbSNP is the largest public resource of polymorphism information. It assembles information from researcher submissions. As of build 128 (October 2007), it contained nearly 12 million unique SNPs for the human genome and close to 34 million more SNPs for 43 different species
International HapMap Project *(79)*	The HapMap project contains genotypes, frequencies, haplotypes, and assay information for phase I and phase II individuals sampled as part of the HapMap project. Public Release #22 (March 2007) contains approximately 3.8 million genotyped SNPs for four different human populations
The Human Gene Mutation Database (HGMD) *(39)*	HGMD is a collection of genetic mutations that underlie or are associated with human inherited disease. As of November 2007, over 57,000 mutations have been curated
Human Genome Variation Database (HGVbase) *(80)*	HGVbase is a curated summary of human DNA variation with a focus on the link between haplotypes and phenotypes. Release 16 holds close to 9 million entries
The ALlele FREquency Database (ALFRED) *(81)*	ALFRED contains allele frequency data from diverse human populations
Online Mendelian Inheritance in Man (OMIM) *(82)*	OMIM is a catalog of human genes and genetic disorders. Information on specific disease-causing mutations is provided with publication cross-references to PubMed
SNPeffect *(28)*	SNPeffect is a database of coding nonsynonymous SNPs combined with information on the likely functional and physiochemical properties of such mutations. Over 133,000 SNPs have been analyzed
SNPper *(83)*	SNPper is a retrieval and SNP analysis database built on top of dbSNP and the UCSC Genome Browser
Japanese Single Nucleotide Polymorphisms database (JSNP) *(84)*	JSNP is a database of common SNPs in the Japanese population. Release 32 (October 2007) contains over 515,000 SNPs genotyped from 934 individuals
dbQSNP *(85)*	dbQSNP is a database of SNPs and associated allele frequencies for polymorphisms in the promoter regions of genes assessed through sequencing and Single-Strand Conformation Polymorphism analysis. Version 14 (October 2007) contains over 10,100 SNPs
The Open Regulatory Annotation database (ORegAnno) *(40)*	ORegAnno is an open-access database of community-curated regulatory regions, transcription factors, and regulatory mutations
CASCAD *(86)*	CASCAD contains candidate SNPs associated with expressed sequences for *Rattus norvegicus* and *Danio rerio*. It supplies enriched annotation to facilitate discrimination of SNPs involved in phenotypic variation of different populations

(continued)

**Table 2
(continued)**

topoSNP *(87)*	topoSNP is a resource for mapping nonsynonymous SNPs to 3D protein structures. Contains integrated nsSNPs from OMIM and dbSNP. Its publication release contains 27,417 nsSNPs corresponding to 770 protein structures
rSNP_Guide *(38)*	rSNP_Guide contains information regarding artificial or natural variation effects on gene expression. The April 2006 release contains 46 entries
SeattleSNPs Variation Discovery Resource *(88)*	The SeattleSNPs Variation Discovery Resource is a project focused on identifying, genotyping, and modeling associations between SNPs that underlie inflammatory response in humans. 318 genes primarily involved in inflammation, lipid metabolism, and blood pressure regulation have been resequenced and over 36,000 SNPs have been found as of November 2007. Genes can be externally nominated to the project for resequencing
GeneSNPs database *(89)*	The NIEHS Environmental Genome Project is a resequencing project for genes involved in disease susceptibility in U.S. populations. As of June 2006, their GeneSNPs database contains close to 83,000 SNPs from 593 genes primarily involved in DNA repair, cell cycle regulation, drug metabolism, and apoptosis
Human Structural Variation Database *(90)*	Database contains large-scale structural variation (LSV), copy number polymorphisms (CNPs), and intermediate-sized structural variation (ISV)
Database of Genomic Variants *(91)*	A curated catalogue of large-scale variation in the human genome
The Chromosome Anomaly Collection *(92)*	Database contains cytogenetically visible mutation

To facilitate access to genetic mutation data, genome databases, like EnsEMBL and the UCSC Genome Browser and protein databases, such as SwissProt *(54)*, have integrated this information within their repositories. This has allowed researchers to seamlessly investigate known mutations in genomic sequences and proteins.

2.3. Derived Alleles and Frequency

An estimate of the age of an allele can be inferred by identifying the ancestral allele in closely related species and calculating a derived allele frequency (DAF; DAFs are the frequency at which the ancestral allele is observed in the reference species population). A DAF can provide information about the ability of an allele under selection to "hitchhike" to low or high frequency *(55)*. Cargill et al. were able to demonstrate that a significant proportion of alleles have risen in frequency since the human–chimpanzee divergence to become the major allele in the population *(56)*. This diagnostic has been used to detect recent positive selection in the immune response genes CAV1 and CAV2 *(57)*.

Furthermore, DAFs have been used to characterize signals of positive selection more generally across the human genome *(17)* and within conserved noncoding sequences *(43)*.

Allele characterization based on ancestral allele status is utilized in tools that prioritize nonsynonymous SNPs such as SIFT *(6)* and PolyPhen *(58)*. Ancestral alleles for humans can be directly obtained with reference to the chimpanzee genome from dbSNP, EnsEMBL, or UCSC Genome Browser.

3. Methods

In this section we will discuss how to predict TFBS using weight matrices, acquire known regulatory elements and sources of variation, and investigate key properties of this variation. We will also investigate two tools; PupaSNP and Sockeye designed for streamlining this analysis in specific genomic intervals.

3.1. Predicting Transcription Factor Binding Sites Using Weight Matrices

There are several online resources for identifying transcription factor binding sites using weight matrices. Here, we provide information on setting up and running your own matrix scanner using the TFBS::Perl modules and the JASPAR database *(31, 49)*. The advantage of this is that both resources are freely available and many sequences can be run simultaneously on a cluster. Furthermore, the TFBS::Perl modules allow key statistics to be easily parsed using built-in parsers.

1. Install the TFBS::Perl modules from http://forkhead.cgb.ki.se/TFBS/

   a. Select the most recent build file and tar and unzip it

   b. Follow the instructions in the README file for installing TFBS::Perl

   c. TFBS::Perl requires both BioPerl and PDL (Perl Data Language) to be installed. BioPerl can be installed from http://www.bioperl.org/wiki/Installing_BioPerl and PDL can be installed from http://pdl.perl.org/

2. Download the JASPAR transcription factor weight matrices from http://jaspar.genereg.net/

   a. The matrices can be downloaded in text or in an SQL-compliant format

   b. To use the SQL-compliant format, create a new database and import the jaspar core data in the tables "matrix_annotation," "matrix_data," and "matrix_info." This can be done in MySQL by issuing the following commands:

```
CREATE DATABASE jaspar;
USE DATABASE jaspar;
LOAD DATA INFILE 'MATRIX_ANNOTATION.txt' INTO
TABLE matrix_annotation;
LOAD DATA INFILE 'MATRIX_DATA.txt' INTO TABLE
matrix_data;
LOAD DATA INFILE 'MATRIX_INFO.txt' INTO TABLE
matrix_info;
```

3. Build a TFBS::Perl script to test a matrix profile(s) of your choice
    a. If using JASPAR via MySQL, connect to it using the following command (substituting the database name, host name, user name, and password as necessary):

```
my $jaspar_db = TFBS::DB::JASPAR4-
>connect("dbi:mysql:jaspar:localhost", "user-
name", "password");
```

   b. If using JASPAR via text files, matrices can be loaded via the command (substituting the path name where the downloaded JASPAR matrices are located as necessary):

```
my $jaspar_db=TFBS::DB::FlatFileDir->connect("/
home/MatrixDir");
```

   c. Using the JASPAR database connection, select matrices for analysis. Matrices can either be assembled into a set based on species or accessed individually using the following commands:

```
my $matrixset = $jaspar_db->get_MatrixSet(-spe-
cies => ['Homo sapiens']);
or,
my $pwm = $jaspar_db->get_Matrix_by_name('NF-
kappaB', 'PWM');
my $matrixset = new TFBS::MatrixSet->new();
$matrixset->add_Matrix($pwm);
```

   d. Set a threshold for analysis and analyze a sequence as a string for a collection of matrices (MatrixSet) using the command:

```
my $siteset = $matrixset ->search_seq( -seqstring
=> $sequence,
-threshold => "80%");
```

   e. The output results are stored to a SiteSet object which contains methods for iterating over the results and reporting the score. Here is an example command to get hits that overlap a position (such as an SNP coordinate):

```
my  $siteset_iterator  =  $siteset->Iterator
(-sort_by =>'score');
while  (my  $site_object  =  $siteset_iterator
->next) {
if ($site_object->start <= $position &&site_
object->end >= $position) {print site_object
->score . "\n";
}
}
```

4. This method for predicting transcription factor binding sites is prone to error when appropriate biological knowledge is not applied. The small number of sites and their proclivity to binding sequences based on random chance means that knowledge of the location of regulatory regions and implicated transcription factors is useful in deriving true transcription factor binding sites. In our own work, we have noticed that unrestricted application of these models is not discriminative when used to identify potential binding sites containing regulatory polymorphisms but begins to improve when information such as coexpression is used to filter the transcription factor matrices that are applied *(42)*

3.2. Acquiring Known Regulatory Elements from ORegAnno

In addition to predicted regulatory elements, a handful of known regulatory regions and transcription factor bindings are available in databases such as ORegAnno *(40)*. The advantage of ORegAnno is that all the data is available either through UCSC, a direct download, or through Web Services. Here are detailed steps for acquiring the regulatory polymorphisms and transcription factor binding sites using either Web Services or the UCSC Genome Browser *(37)*.

1. To obtain the ORegAnno data from the UCSC Genome Browser

   a. Select the table browser link from the UCSC Genome Browser

   b. Select the "Expression and Regulation" track for the genome and assembly of interest

   c. A smaller interval can be selected using the "region" filter

   d. Select the desired "output format"

   e. Run "get output." This will fetch the ORegAnno data stored in the UCSC Genome Browser. However, this information only contains the stable ids of the ORegAnno entries. To obtain the "type" of entry, reselect the table of "oregannoAttr"

2. To obtain the ORegAnno data from using a SOAP::Lite Perl script:

   a. If not installed, install the SOAP::Lite modules from http://www.cpan.org

b. Examples of scripts and methods for fetching ORegAnno data can be found at http://www.oreganno.org/oregano/ Dump.jsp

c. The advantage of this method is that subsets of the data can be selected based on keywords in their abstract or identifier fields, such as "cancer" or "Sp1"

d. Declare a remote connection to ORegAnno using the command:

```
my $osa = SOAP::Lite
-> uri('OregannoServerImpl')
->    proxy('http://www.oreganno.org/oregano/
soap/');
```

e. Acquire data for known regulatory polymorphisms from ORegAnno using the command:

```
my $response = $osa->searchRecords(SOAP::Data
->name(field   =>   SOAP::Data->type(string   =>
"type")),
SOAP::Data->name(query=>SOAP::Data->type(string
=> "REGULATORY POLYMORPHISM")));
```

f. Similarly, data for known transcription factor binding sites can be obtained using the command:

```
my $response = $osa->searchRecords(SOAP::Data->
name(field   =>   SOAP::Data->type(string   =>
"type")),
SOAP::Data->name(query=>SOAP::Data->type(string
=> "TRANSCRIPTION FACTOR BINDING SITE")));
```

g. Several fields are available for searching in ORegAnno. These fields are the same as those available from the Search page at http://www.oreganno.org/oregano/Search.jsp

3.3. Acquiring Sources of Single Nucleotide Polymorphisms

In this section, we will describe how to access en masse genetic variation from dbSNP *(59)*, EnsEMBL *(36)*, and UCSC *(37)*. We will also discuss how to download genotype information from the HapMap Consortium.

1. Downloading genetic variation from dbSNP:

   a. SNPs can be downloaded by organism and chromosome using the public FTP site of dbSNP at ftp://ftp.ncbi.nih. gov/snp/organisms/ (*see* **Note 1**)

   b. Alternately, to programmatically acquire the SNP data on an SNP by SNP basis, the World Wide Web library for Perl (LWP) and NCBI's eutils can be used

   c. To access data in XML format using eutils and LWP, one can issue the following command (where id is the rs number such that "rs1000" becomes an id of "1000"):

```
my $browser_results = LWP::UserAgent->new;
my $url_results = "http://eutils.ncbi.nlm.nih.
gov/entrez/eutils/efetch.fcgi?db=snp&id=".
$id."&report=XML";
my $response = $browser_results->get( $url_
results );
```

    d. Using any XML parser will allow information from this dbSNP record to be extracted

2. Downloading genetic variation from EnsEMBL:

    a. EnsEMBL also provides a URL-based mechanism of viewing SNP data or a programmatic method for accessing the data en masse

    b. SNP data can be viewed by looking at the SNPView URL for the selected species. This can be accessed by entering an "rs" number into the search field

    c. To download SNP data, on cane either use the BioMart toolkit embedded within EnsEMBL at http://www. ensembl.org/biomart/martview/ or use the EnsEMBL Perl API

    d. To use BioMart, select "SNP" from the "Choose Database" field. A Dataset such as "Human" can then be selected. By clicking on "Filters," specific genomic intervals and selection criteria can be defined for extraction. By clicking on "Attributes," various output attribute data can be added to the output, such as coordinates, alleles, and sequences. Once the "Filters" and "Attributes" have been selected, the "Results" button processes the request and displays the information

    e. To use the EnsEMBL Perl API, it must first be installed following the instructions at http://www.ensembl.org/ info/software/api_installation.html for both the "core" and "variation" database APIs. Once installed, the following commands can be run to initiate database adaptors and extract SNP data: (*see* **Note 2**)

```
Bio::EnsEMBL::Registry->load_all();
my $sa = Bio::EnsEMBL::Registry->get_adaptor
("human", "core", "Slice");
my $vfa = Bio::EnsEMBL::Registry->get_adaptor
("human", "variation", "VariationFeature");
my $slice = $sa->fetch_by_region('chromosome',
'X', 1e6, 2e6);
my @variation_features = @{$vfa->fetch_all_by_
Slice($slice)};
```

3. Downloading genetic variation data from the UCSC Genome Browser

a. Variation data can be downloaded from the UCSC Genome Browser using their table browser or directly from their download site at http://hgdownload.cse.ucsc.edu/downloads.html under "Annotation Database"

b. To download via the table browser, enter the table browser, by clicking on "Tables" on the UCSC Genome Browser site

c. In the "Group" field, select "Variation and Repeats" and select the associated "Track" for output

d. After filters and output formats have been entered, the "get output" button returns the associated variation data

4. Extracting genotype data from the HapMap

a. All of the above databases contain HapMap data for download; however, the current information can also easily be extracted from the HapMap site at www.hapmap.org using either HapMart or the "Bulk Data Download"

b. To obtain the current variants and the population genotypes from the Bulk Data Download section, click on "Genotypes" under the "Bulk Data" tab

c. Follow the links to the latest, nonredundant source of information

d. The output provides the observed alleles for each individual in the sampled population

e. Other information such as recombination rates and hotspots derived from the HapMap data can also be downloaded from this site

3.4. Calculating Ancestral Alleles and Derived Allele Frequency

When analyzing SNP data, the "ancestral" allele is defined as the sequence found in the last common ancestor of two closely related species. The "derived" allele is identified as the newly arisen sequence. Identifying the accurate ancestral allele is important for assessing evolutionary properties of the variant within its descendent species. In this section, we will discuss a method of calculating the ancestral allele and its associated derived allele frequency.

1. The ancestral allele as provided by dbSNP and imported in EnsEMBL can be obtained using the database access methods in the previous section. This method relies on a comparison of human DNA to chimpanzee DNA *(60)*. We will discuss a method using the EnsEMBL Perl API for calculating the ancestral allele from alignments of human and chimpanzee. This method is discussed since it can be ubiquitously applied to any pair of species in EnsEMBL when attempting to determine the ancestral allele. However, similar analyses can be conducted using whole genome alignments from any database

2. Given a human SNP coordinate in an EnsEMBL imported genome assembly, connect to both the human SliceAdaptor

from the EnsEMBL "core" database and the DnaAlignAdaptor from the EnsEMBL "compara" database. This can be done using the following commands: (*see* **Note 2**)

```
Bio::EnsEMBL::Registry->load_all();
my $sa = Bio::EnsEMBL::Registry->get_adaptor
("human", "core", "Slice");
my $dafa = Bio::EnsEMBL::Registry->get_adaptor
("compara",'compara','DnaAlignFeature');
```

3. Obtain a genomic slice for the given SNP coordinates

```
my $slice = $sa->fetch_by_region('chromosome',
$chromosome, $start, $end, $strand );
```

4. Obtain alignments from BLASTZ *(61)* for chimpanzee that overlap the genomic slice using the commands: (*see* **Note 3**)

```
my @chimp_dna_align_features = @{$dafa->fetch_
all_by_Slice($slice, 'Pan troglodytes', undef,
'BLASTZ_NET')};
```

5. These alignments are then restricted to the interval of the SNP from which the corresponding sequence can be extracted with the following commands:

```
my $restricted_feature = $chimp_dna_align_fea-
ture->restrict_between_positions($slice->start,
$slice->end, "SEQ");
my $sequence = @{$restricted_feature->alignment_
strings}[1];
```

6. The ancestral allele is identified as the allele that is common between chimpanzee and the human allele (**Fig. 2**). It is important

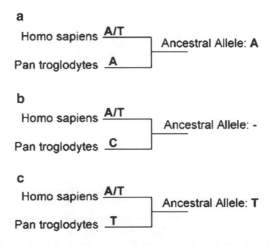

Fig. 2. Calculating ancestral alleles. Ancestral alleles are calculated based on similarity with the most recent common ancestor. In these examples, we assume that an A/T polymorphism has been identified in a human population. In (**a**) and (**c**), the chimpanzee orthologous base matches one of the two human alleles; this match defines the ancestral allele. In (**b**), there is no match which can be made and the ancestral allele is undefined.

to only do this comparison between very closely related species so as to avoid potential frequency biases by selecting only on well-conserved sequences

3.5. Interpreting the Relationships between Predicted Transcription Factor Binding Sites and Regulatory Variants

In the previous sections, we identified methods for acquiring predictions of transcription factor binding sites and single nucleotide polymorphisms; in this section, a method of combining the two is described for predicting which variants may affect regulatory function. However, there are two important caveats when using transcription factor binding matrix-derived predictions for regulatory SNP analysis which may considerably influence the number of type I errors that are made. The first caveat is that the more *a priori* biological knowledge you have of which transcription factors are present, the fewer spurious transcription factor binding site predictions you will make. The second caveat, in a similar notion, is that *a priori* knowledge of the existence of an investigated variant within a regulatory region and its proximity to the TSS also significantly influences the quality of predictions that can be made. All things considered, these two factors are likely to make a large difference in the accuracy of your analysis.

1. Choose a high stringency threshold for TFBS predictions (>=80%). The justification for this is that we will be assessing how a prediction changes in an allele-specific manner. A low stringency biases the results toward changes which are only reflective of loss of information content in a specific binding profile position.

2. Calculate predictions for each allele with sufficient flank to identify the longest matrix on each side of the variant's position. For example, if the longest matrix was 11 bp, sufficient sequence length would be 21 bases with the variant in the middle. However, it is not important how long the sequence is as long as it is not too short.

3. Identify predictions overlapping the variant where one of the alleles meets the stringency threshold.

4. A differential score can be obtained for each allele's prediction. A comparable way of assessing this is to quantify the magnitude of the threshold drop in the TFBS binding predictions (**Fig. 3**).

5. The importance of each SNP can be ranked as a function of the differential score and other information such as allele frequency and functional context (including position in the genome, conservation status and correlation with other functional information, such as expression, regulatory element, and differential binding annotation).

3.6. Using PupaSuite and Sockeye for Regulatory Variation Analysis

Integrated tools have become available to allow researchers to perform rudimentary analyses to prioritize potential regulatory polymorphisms. In this section, we will describe methods for using PupaSuite *(28)* to perform gene-centric analysis of putative

atcgatcggTATAAAgcagttcgac   atcgatcggTACAAAgcagttcgac
Allele 1: **T**   Allele 2: **C**
Overlapping Prediction Score: **99%**   Overlapping Prediction Score: **75%**

Differential Score = abs(Overlapping Prediction Score (allele 1) - Overlapping Prediction Score (allele 2))
= abs(**99 - 75**)
= **14**

Fig. 3. Calculating an allele-specific differential score. A TATA-box binding profile is applied to each sequence with the tested alleles in situ and the prediction score is calculated. Here, one of the alleles exceeds the 80% threshold. The absolute value of the difference in this score is calculated.

regulatory polymorphisms and a method of using a tool called Sockeye *(36)* for performing targeted TFBS prediction in the context of genetic variation embedded in EnsEMBL.

3.6.1. Using PupaSuite

1. PupaSuite is located online at http://pupasuite.bioinfo.cipf.es/.

2. Select a list of genes, a genomic location, a list of SNPs, or an input haplotype for analysis by clicking on the appropriate tab. For each gene, only the region 10 kb upstream is analyzed.

3. Select either TRANSFAC or JASPAR as a source for TFBS predictions. Note: PupaSuite looks for co-occurrence of high-scoring TFBS predictions with SNPs and not allele-specific changes in scores as described in the previous section.

4. Optionally select whether to only include predictions within human–mouse conserved regions or to include all regions.

5. Click "Run" and a list of SNPs overlapping high-scoring TFBS are output.

3.6.2. Using Sockeye

1. Sockeye is available as a Java application from http://www. bcgsc.ca/sockeye/ It can be run by selecting the Java Web Start link. (*see* **Note 4**)

2. When Sockeye starts, an empty genomic interval is displayed as a blue plane. A genomic interval can be selected by clicking on "Data" and then "Query Data/Import Data…" and selecting an EnsEMBL database location and loading a genomic coordinate.

3. Genes and variation can be displayed by toggling the appropriate field in the lower-left window.

4. Regions can be selected by right clicking and moving across the blue plane. These regions can be used to dynamically execute various types of transcription factor binding site predictions. Alternately by clicking on a gene, an orthologous gene track can be loaded and information regarding conservation can be displayed.

4. Notes

1. dbSNP records single nucleotide polymorphisms using ids called "ss" numbers and "rs" numbers. The "rs" numbers indicate that the variant is non-redundant whereas the "ss" number indicates a submitted SNP.

2. The EnsEMBL Registry points to the file defined by the environment variable $ENSEMBL_REGISTRY when using the registry method to connect to the EnsEMBL database.

3. Other alignment methods are also available from the EnsEMBL compara database. These can be listed by viewing the contents of the "analysis" database.

4. The current version of Sockeye supports an outdated version of EnsEMBL and must be edited manually or used with GFF files downloaded from UCSC or EnsEMBL.

Acknowledgment

S.B.M. would like to thank Monica C. Sleumer, Daniel C. Jeffares, and Emmanouil T. Dermitzakis for critical review and support in development of this work. S.B.M. is funded by the European Molecular Biology Organization and the Natural Sciences and Engineering Research Council of Canada.

References

1. Pastinen T, Hudson TJ (2004) Cis-acting regulatory variation in the human genome. Science 306: 647–650.

2. Knight JC (2005) Regulatory polymorphisms underlying complex disease traits. J Mol Med 83: 97–109.

3. Rockman MV, Wray GA (2002) Abundant raw material for cis-regulatory evolution in humans. Mol Biol Evol 19: 1991–2004.

4. Wittkopp PJ (2005) Genomic sources of regulatory variation in cis and in trans. Cell Mol Life Sci 62: 1779–1783.

5. Whitehead A, Crawford DL (2006) Variation within and among species in gene expression: raw material for evolution. Mol Ecol 15: 1197–1211.

6. Miao X, Yu C, Tan W, Xiong P, Liang G, et al. (2003) A functional polymorphism in the matrix metalloproteinase-2 gene promoter (-1306C/T) is associated with risk of development but not metastasis of gastric cardia adenocarcinoma. Cancer Res 63: 3987–3990.

7. Bond GL, Hu W, Bond EE, Robins H, Lutzker SG, et al. (2004) A single nucleotide polymorphism in the MDM2 promoter attenuates the p53 tumor suppressor pathway and accelerates tumor formation in humans. Cell 119: 591–602.

8. Caspi A, Sugden K, Moffitt TE, Taylor A, Craig IW, et al. (2003) Influence of life stress on depression: moderation by a polymorphism in the 5-HTT gene. Science 301: 386–389.

9. Prokunina L, Castillejo-Lopez C, Oberg F, Gunnarsson I, Berg L, et al. (2002) A regulatory polymorphism in PDCD1 is associated with susceptibility to systemic lupus erythematosus in humans. Nat Genet 32: 666–669.

10. Kostrikis LG, Neumann AU, Thomson B, Korber BT, McHardy P, et al. (1999) A polymorphism in the regulatory region of the CC-chemokine receptor 5 gene influences perinatal transmission of human immunodeficiency virus type 1 to African-American infants. J Virol 73: 10264–10271.

11. Saito H, Tada S, Ebinuma H, Wakabayashi K, Takagi T, et al. (2001) Interferon regulatory factor 1 promoter polymorphism and response to type 1 interferon. J Cell Biochem Suppl 36: 191–200.

12. Stranger BE, Nica AC, Forrest MS, Dimas A, Bird CP, et al. (2007) Population genomics of human gene expression. Nat Genet 39: 1217–1224.

13. Pastinen T, Sladek R, Gurd S, Sammak A, Ge B, et al. (2004) A survey of genetic and epigenetic variation affecting human gene expression. Physiol Genomics 16: 184–193.

14. Frazer KA, Ballinger DG, Cox DR, Hinds DA, Stuve LL, et al. (2007) A second generation human haplotype map of over 3.1 million SNPs. Nature 449: 851–861.

15. Nielsen R, Hellmann I, Hubisz M, Bustamante C, Clark AG (2007) Recent and ongoing selection in the human genome. Nat Rev Genet 8: 857–868.

16. Sabeti PC, Reich DE, Higgins JM, Levine HZ, Richter DJ, et al. (2002) Detecting recent positive selection in the human genome from haplotype structure. Nature 419: 832–837.

17. Voight BF, Kudaravalli S, Wen X, Pritchard JK (2006) A map of recent positive selection in the human genome. PLoS Biol 4: e72.

18. Sabeti PC, Varilly P, Fry B, Lohmueller J, Hostetter E, et al. (2007) Genome-wide detection and characterization of positive selection in human populations. Nature 449: 913–918.

19. King MC, Wilson AC (1975) Evolution at two levels in humans and chimpanzees. Science 188: 107–116.

20. Kornblihtt AR (2005) Promoter usage and alternative splicing. Curr Opin Cell Biol 17: 262–268.

21. Davidson EH (2001) Genomic Regulatory Systems: Development and Evolution. San Diego: Academic. xii, 261 pp.

22. Hoogendoorn B, Coleman SL, Guy CA, Smith K, Bowen T, et al. (2003) Functional analysis of human promoter polymorphisms. Hum Mol Genet 12: 2249–2254.

23. Hewett D, Lynch J, Child A, Firth H, Sykes B (1994) Differential allelic expression of a fibrillin gene (FBN1) in patients with Marfan syndrome. Am J Hum Genet 55: 447–452.

24. Morley M, Molony CM, Weber TM, Devlin JL, Ewens KG, et al. (2004) Genetic analysis of genome-wide variation in human gene expression. Nature 430: 743–747.

25. Monks SA, Leonardson A, Zhu H, Cundiff P, Pietrusiak P, et al. (2004) Genetic inheritance of gene expression in human cell lines. Am J Hum Genet 75: 1094–1105.

26. Cheung VG, Spielman RS, Ewens KG, Weber TM, Morley M, et al. (2005) Mapping determinants of human gene expression by regional and genome-wide association. Nature 437: 1365–1369.

27. Pastinen T, Ge B, Hudson TJ (2006) Influence of human genome polymorphism on gene expression. Hum Mol Genet 15 Spec No 1: R9–R16.

28. Conde L, Vaquerizas JM, Dopazo H, Arbiza L, Reumers J, et al. (2006) PupaSuite: finding functional single nucleotide polymorphisms for large-scale genotyping purposes. Nucleic Acids Res 34: W621–W625.

29. Freimuth RR, Stormo GD, McLeod HL (2005) PolyMAPr: programs for polymorphism database mining, annotation, and functional analysis. Hum Mutat 25: 110–117.

30. Matys V, Kel-Margoulis OV, Fricke E, Liebich I, Land S, et al. (2006) TRANSFAC and its module TRANSCompel: transcriptional gene regulation in eukaryotes. Nucleic Acids Res 34: D108–D110.

31. Bryne JC, Valen E, Tang MH, Marstrand T, Winther O, et al. (2008) JASPAR, the open access database of transcription factor-binding profiles: new content and tools in the 2008 update. Nucleic Acids Res 36: D102–D106.

32. Tomso DJ, Inga A, Menendez D, Pittman GS, Campbell MR, et al. (2005) Functionally distinct polymorphic sequences in the human genome that are targets for p53 transactivation. Proc Natl Acad Sci USA 102: 6431–6436.

33. Mottagui-Tabar S, Faghihi MA, Mizuno Y, Engstrom PG, Lenhard B, et al. (2005) Identification of functional SNPs in the 5-prime flanking sequences of human genes. BMC Genomics 6: 18.

34. Khan IA, Mort M, Buckland PR, O'Donovan MC, Cooper DN, et al. (2005) In silico discrimination of single nucleotide polymorphisms and pathological mutations in human gene promoter regions by means of local DNA sequence context and regularity. In Silico Biol 6: 0003.

35. Mooney SD, Altman RB (2003) MutDB: annotating human variation with functionally relevant data. Bioinformatics 19: 1858–1860.

36. Montgomery SB, Astakhova T, Bilenky M, Birney E, Fu T, et al. (2004) Sockeye: a 3D

environment for comparative genomics. Genome Res 14: 956–962.

37. Hinrichs AS, Karolchik D, Baertsch R, Barber GP, Bejerano G, et al. (2006) The UCSC Genome Browser Database: update 2006. Nucleic Acids Res 34: D590–D598.

38. Ponomarenko JV, Merkulova TI, Orlova GV, Fokin ON, Gorshkova EV, et al. (2003) rSNP_Guide, a database system for analysis of transcription factor binding to DNA with variations: application to genome annotation. Nucleic Acids Res 31: 118–121.

39. Stenson PD, Ball EV, Mort M, Phillips AD, Shiel JA, et al. (2003) Human Gene Mutation Database (HGMD): 2003 update. Hum Mutat 21: 577–581.

40. Griffith OL, Montgomery SB, Bernier B, Chu B, Kasaian K, et al. (2008) ORegAnno: an open-access community-driven resource for regulatory annotation. Nucleic Acids Res 36: D107–D113.

41. Portales-Casamar E, Kirov S, Lim J, Lithwick S, Swanson MI, et al. (2007) PAZAR: a framework for collection and dissemination of cis-regulatory sequence annotation. Genome Biol 8: R207.

42. Montgomery SB, Griffith OL, Schuetz JM, Brooks-Wilson A, Jones SJ (2007) A survey of genomic properties for the detection of regulatory polymorphisms. PLoS Comput Biol 3: e106.

43. Drake JA, Bird C, Nemesh J, Thomas DJ, Newton-Cheh C, et al. (2006) Conserved noncoding sequences are selectively constrained and not mutation cold spots. Nat Genet 38: 223–227.

44. Khaitovich P, Paabo S, Weiss G (2005) Toward a neutral evolutionary model of gene expression. Genetics 170: 929–939.

45. Balhoff JP, Wray GA (2005) Evolutionary analysis of the well characterized endo16 promoter reveals substantial variation within functional sites. Proc Natl Acad Sci USA 102: 8591–8596.

46. Romano LA, Wray GA (2003) Conservation of Endo16 expression in sea urchins despite evolutionary divergence in both cis and trans-acting components of transcriptional regulation. Development 130: 4187–4199.

47. Klug SJ, Famulok M (1994) All you wanted to know about SELEX. Mol Biol Rep 20: 97–107.

48. Stormo GD, Schneider TD, Gold L, Ehrenfeucht A (1982) Use of the 'Perceptron' algorithm to distinguish translational initiation sites in E. coli. Nucleic Acids Res 10: 2997–3011.

49. Lenhard B, Wasserman WW (2002) TFBS: computational framework for transcription factor binding site analysis. Bioinformatics 18: 1135–1136.

50. Crooks GE, Hon G, Chandonia JM, Brenner SE (2004) WebLogo: a sequence logo generator. Genome Res 14: 1188–1190.

51. Base BCMD (2006) http://research.nhgri.nih.gov/bic/

52. Clifford R, Edmonson M, Hu Y, Nguyen C, Scherpbier T, et al. (2000) Expression-based genetic/physical maps of single-nucleotide polymorphisms identified by the cancer genome anatomy project. Genome Res 10: 1259–1265.

53. CFMDB (2006) http://www.genet.sickkids.on.ca/cftr/

54. Wu CH, Apweiler R, Bairoch A, Natale DA, Barker WC, et al. (2006) The Universal Protein Resource (UniProt): an expanding universe of protein information. Nucleic Acids Res 34: D187–D191.

55. Fay JC, Wu CI (2000) Hitchhiking under positive Darwinian selection. Genetics 155: 1405–1413.

56. Cargill M, Altshuler D, Ireland J, Sklar P, Ardlie K, et al. (1999) Characterization of single-nucleotide polymorphisms in coding regions of human genes. Nat Genet 22: 231–238.

57. Walsh EC, Sabeti P, Hutcheson HB, Fry B, Schaffner SF, et al. (2006) Searching for signals of evolutionary selection in 168 genes related to immune function. Hum Genet 119: 92–102.

58. Ramensky V, Bork P, Sunyaev S (2002) Human non-synonymous SNPs: server and survey. Nucleic Acids Res 30: 3894–3900.

59. Sherry ST, Ward MH, Kholodov M, Baker J, Phan L, et al. (2001) dbSNP: the NCBI database of genetic variation. Nucleic Acids Res 29: 308–311.

60. Spencer CC, Deloukas P, Hunt S, Mullikin J, Myers S, et al. (2006) The influence of recombination on human genetic diversity. PLoS Genet 2: e148.

61. Schwartz S, Kent WJ, Smit A, Zhang Z, Baertsch R, et al. (2003) Human-mouse alignments with BLASTZ. Genome Res 13: 103–107.

62. Palaniswamy SK, James S, Sun H, Lamb RS, Davuluri RV, et al. (2006) AGRIS and AtRegNet. a platform to link cis-regulatory elements and transcription factors into regulatory networks. Plant Physiol 140: 818–829.

63. AtProbe (2006) http://rulai.cshl.edu/software/index1.htm

64. CEPDB (2006) http://rulai.cshl.edu/software/index1.htm

65. Kummerfeld SK, Teichmann SA (2006) DBD: a transcription factor prediction database. Nucleic Acids Res 34: D74–D81.

66. Barrasa MI, Vaglio P, Cavasino F, Jacotot L, Walhout AJ (2007) EDGEdb: a transcription factor-DNA interaction database for the

analysis of *C. elegans* differential gene expression. BMC Genomics 8: 21.

67. Schmid CD, Perier R, Praz V, Bucher P (2006) EPD in its twentieth year: towards complete promoter coverage of selected model organisms. Nucleic Acids Res 34: D82–D85.

68. Adryan B, Teichmann SA (2006) FlyTF: a systematic review of site-specific transcription factors in the fruit fly *Drosophila melanogaster*. Bioinformatics 22: 1532–1533.

69. Pohar TT, Sun H, Davuluri RV (2004) HemoPDB: Hematopoiesis Promoter Database, an information resource of transcriptional regulation in blood cell development. Nucleic Acids Res 32: D86–D90.

70. LSPD (2006) http://rulai.cshl.edu/software/index1.htm

71. Palaniswamy SK, Jin VX, Sun H, Davuluri RV (2005) OMGProm: a database of orthologous mammalian gene promoters. Bioinformatics 21: 835–836.

72. Grienberg I, Benayahu D (2005) Osteo-Promoter Database (OPD) – promoter analysis in skeletal cells. BMC Genomics 6: 46.

73. Higo K, Ugawa Y, Iwamoto M, Korenaga T (1999) Plant cis-acting regulatory DNA elements (PLACE) database: 1999. Nucleic Acids Res 27: 297–300.

74. Lescot M, Dehais P, Thijs G, Marchal K, Moreau Y, et al. (2002) PlantCARE, a database of plant cis-acting regulatory elements and a portal to tools for in silico analysis of promoter sequences. Nucleic Acids Res 30: 325–327.

75. Shahmuradov IA, Gammerman AJ, Hancock JM, Bramley PM, Solovyev VV (2003) PlantProm: a database of plant promoter sequences. Nucleic Acids Res 31: 114–117.

76. Zhu J, Zhang MQ (1999) SCPD: a promoter database of the yeast *Saccharomyces cerevisiae*. Bioinformatics 15: 607–611.

77. Kanamori M, Konno H, Osato N, Kawai J, Hayashizaki Y, et al. (2004) A genome-wide and nonredundant mouse transcription factor database. Biochem Biophys Res Commun 322: 787–793.

78. Kolchanov NA, Ignatieva EV, Ananko EA, Podkolodnaya OA, Stepanenko IL, et al. (2002) Transcription Regulatory Regions Database (TRRD): its status in 2002. Nucleic Acids Res 30: 312–317.

79. The HapMap Consortium (2005) A haplotype map of the human genome. Nature 437: 1299–1320.

80. Fredman D, Munns G, Rios D, Sjoholm F, Siegfried M, et al. (2004) HGVbase: a curated resource describing human DNA variation and phenotype relationships. Nucleic Acids Res 32: D516–D519.

81. Rajeevan H, Osier MV, Cheung KH, Deng H, Druskin L, et al. (2003) ALFRED: the ALelle FREquency Database. Update. Nucleic Acids Res 31: 270–271.

82. OMIM (2006) Online Mendelian Inheritance in Man, OMIM (TM). McKusick-Nathans Institute for Genetic Medicine, Johns Hopkins University (Baltimore, MD) and National Center for Biotechnology Information, National Library of Medicine (Bethesda, MD), June 2006. World Wide Web URL: http://www.ncbi.nlm.nih.gov/omim/.

83. Riva A, Kohane IS (2002) SNPper: retrieval and analysis of human SNPs. Bioinformatics 18: 1681–1685.

84. Hirakawa M, Tanaka T, Hashimoto Y, Kuroda M, Takagi T, et al. (2002) JSNP: a database of common gene variations in the Japanese population. Nucleic Acids Res 30: 158–162.

85. Tahira T, Baba S, Higasa K, Kukita Y, Suzuki Y, et al. (2005) dbQSNP: a database of SNPs in human promoter regions with allele frequency information determined by single-strand conformation polymorphism-based methods. Hum Mutat 26: 69–77.

86. Guryev V, Berezikov E, Cuppen E (2005) CASCAD: a database of annotated candidate single nucleotide polymorphisms associated with expressed sequences. BMC Genomics 6: 10.

87. Stitziel NO, Binkowski TA, Tseng YY, Kasif S, Liang J (2004) topoSNP: a topographic database of non-synonymous single nucleotide polymorphisms with and without known disease association. Nucleic Acids Res 32: D520–D522.

88. SeattleSNPs (2006) NHLBI Program for Genomic Applications, SeattleSNPs, Seattle, WA (URL: http://pga.gs.washington.edu) [Accessed 30 Jul 2006].

89. GeneSNPs (2006) NIEHS SNPs. NIEHS Environmental Genome Project, University of Washington, Seattle, WA (URL: http://egp.gs.washington.edu) [Accessed 30 Jul 2006].

90. HSVD (2006) http://humanparalogy.gs.washington.edu/structuralvariation/

91. Iafrate AJ, Feuk L, Rivera MN, Listewnik ML, Donahoe PK, et al. (2004) Detection of large-scale variation in the human genome. Nat Genet 36: 949–951.

92. Barber JC (2005) Directly transmitted unbalanced chromosome abnormalities and euchromatic variants. J Med Genet 42: 609–629.

Chapter 6

Methods of Information Geometry in Computational System Biology (Consistency between Chemical and Biological Evolution)

Vadim Astakhov

Summary

Interest in simulation of large-scale metabolic networks, species development, and genesis of various diseases requires new simulation techniques to accommodate the high complexity of realistic biological networks. Information geometry and topological formalisms are proposed to analyze information processes. We analyze the complexity of large-scale biological networks as well as transition of the system functionality due to modification in the system architecture, system environment, and system components.

The dynamic core model is developed. The term *dynamic core* is used to define a set of causally related network functions. *Delocalization of dynamic core* model provides a mathematical formalism to analyze migration of specific functions in biosystems which undergo structure transition induced by the environment. The term *delocalization* is used to describe these processes of migration. We constructed a holographic model with self-poetic *dynamic cores* which preserves functional properties under those transitions. Topological constraints such as Ricci flow and Pfaff dimension were found for statistical manifolds which represent biological networks. These constraints can provide insight on processes of degeneration and recovery which take place in large-scale networks. We would like to suggest that therapies which are able to effectively implement estimated constraints, will successfully adjust biological systems and recover altered functionality. Also, we mathematically formulate the hypothesis that there is a direct consistency between biological and chemical evolution. Any set of causal relations within a biological network has its dual reimplementation in the chemistry of the system environment.

Key words: System biology, Metabolic network, Biological networks, Information geometry, Complexity, Dynamic systems

1. Introduction

Modern research in synthetic biology and large-scale simulations of biological networks raise many mathematical questions about modeling of the complex networks in silico. These questions

Vadim Astakhov (ed.), *Biomedical Informatics,* Methods in Molecular Biology, vol. 569
DOI 10.1007/978-1-59745-524-4_6, © Humana Press, a part of Springer Science+Business Media, LLC 2009

cover the area of metabolic pathways, protein–protein interactions, gene networks, and chemical therapy. The topic of signaling pathway evolution and multiple implementations has become a subject of scientific research which emerges as an attempt to restore vital functions altered due to a disease or environment changes. We assume that vital biological functions consist entirely of physiological activity in the tissue which leads to the suggestion that complex biological functions emerged from complex biological signaling network architectures. Thus, alteration and recovery processes for such networks can be simulated with algorithms which capture the complexity of the unimpaired biological networks.

One way to analyze complex metabolic networks is with the matrix model *(1)* where the matrix elements represent fluxes of chemical reactions involved in metabolic pathways and boundary conditions represent balances of mass, energy, and solvent capacity. Such formalism implies Euclidean geometry for predefined vectors. We are interested in generalizing matrix formalism to differential geometry defined on statistical manifolds.

Also, we propose a new process called *delocalization* for multiple implementation of predefined function which is not a direct copying of the network features but rather reestablishing new functionality due to adjustment in the system.

An illustration of *delocalization* can be taken from biosynthesis of NAD(P) protein in bacterial pathogens and related species *(2)*. As demonstrated in *(2)*, there are at least six different but entangled pathways to synthesize NAD. Each pathway takes place in certain bacteria but they all share common blocks.

Consider another example **(Fig. 1)** from network theory: node A sends a message to node C through the shortest path

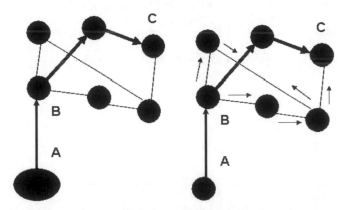

Fig. 1. The *left figure* represents smart-node architecture for communication nodes where a smart node A sends a request to pass a message to node C through known shortest path from A to C. The same functionality can be implemented (*right*) by broadcasting the message to all neighbors.

on the network. That case can be found in many biological systems. The "smart" node A should calculate the shortest paths among network nodes and send a message to a proper neighbor. An arbitrary node should be able to perform a few basic operations which we have schematically defined as: *get message*, *find next in the path*, and *send message to the next*. As can be easily seen, the system is not functioning if the ability to find the shortest path is destroyed. But likely, similar functionality can be recovered effectively if each node can be adjusted to perform just simple operations: *get message* and *send message to all neighbors*. The signal will be transmitted from A to C through many paths as well as through the shortest path. We can say that *find the shortest path* function has implicitly emerged in this model as opposed to explicitly implemented in the smart node model.

Ability to broadcast messages is crucial here. For example, if a message is not broadcast but sent to a random node, that will not lead to emergence of the *shortest path* functionality. This is obviously a very noisy example but it illustrates the possibility of recovering the lost functionality by modification in the network architecture.

The whole idea of *delocalization* is a functional analog of the holography effect which is well known in physics. Aside from an optical application there is a mathematical formalism which provides a basis for the holographic representation of a physical system. A physical system can be described by a set of descriptors localized in $d + 1$-dimensional space. Also, the system can be described by its holographic representation in d-dimensional space. This holographic representation requires none-local descriptors which are spatially distributed across d-dimensional space. The general formalism from mathematical physics is known as the holographic principle conjuncture *(3)*. This is a correspondence between a system localized in Anti-de Sitter (AdS) space and the system holographic representation in terms of Conformal Field Theory (CFT) *(3)*. It was proposed by Hooft and demonstrated by Maldacena, Witten, and others that certain conformal field theories in d-dimensions can be described on the product of $d + 1$-dimensional AdS space with a compact manifold. Holographic conjuncture was heavily discussed in physics recently. But we are putting aside its original physical context and going to employ mathematical formalism for statistical manifolds which represent the functionality of biological systems. That formalism is used to build a duel-model of causal relations taking place in biosystems. We are going to utilize the fact that delocalized holographic representation preserves all causal relations of the original-localized system.

Methods from the theory of *dynamical systems* are employed to provide geometric formalism for analysis of network systems. A *dynamical system* is a mathematics concept in which a fixed rule describes the time dependence of a point in a geometric space. A *system state* is determined by a collection of numbers that can be measured. Small changes in the state of the system correspond to small changes in the numbers. We use the concept of *informational manifold* from "information and statistical geometry" *(4)* as the geometric *space* and introduce a *system state* as a point on the manifold to describe the dynamics of processes. The numbers are also the coordinates of a geometric space – a manifold. The *evolution* of the dynamical system is a rule that describes what and how future states follow from the current state.

As a first step, we define a statistical manifold to represent biosystem functionality in geometric terms. Geometric formalism naturally merges with AdS formalism and CFT Yang–Mills formalism used in the work of Witten and others *(3)*. Then, we define system dynamics in terms of $d + 1$-dimensional AdS model defined on our manifold. At the same time we consider a holographic representation of the same system in terms of d-dimensional CFT model defined on the conformal boundary of this manifold. Migration to the holographic representation can be modeled as a geometric evolution or geometric flow *(5)* which deforms the statistical manifold metric in a manner formally analogous to the diffusion of heat, thereby smoothing out irregularities in the metric.

Finally, we suggest a Holographic Renormalization Group Flow *(6)* as an algorithm for holographic reconstruction of the system functions. Based on the analysis of the RG-flow *(6, 7)*, we demonstrate that Ricci flow *(8, 9)* preserves causal relations on statistical manifolds. That result can be interpreted for a particular therapy as a set of functional constraints. If those constraints will be implemented by a therapy, then the modified system will be functionally equivalent to the unimpaired biological system.

2. Information Geometry for Analysis of Dynamic Biosystem

We consider a biological network as a dynamic system X composed of n units $\{x_i\}$. Each unit can represent an object like a single neuron or a subnet of the brain network. Those units can be either "on" or "off" with some probability. "On" means an element contributes to the activity that leads to emergence of the mental function Q and "off" is otherwise. Thus the observable state of the function $Q = (Q_1, Q_2,...)$ for the system X can be characterized by certain sets of statistical parameters $(x_1, x_2,...)$ with given probability distribution $p(X|Q)$. This distribution provides causal information about elements involved in the system vital dynamics.

The idea of endowing the space of such parameters with metric and geometric structure leads to the proposal to use Fisher information as a metric of geometric space for $p(X|Q)$-distributions:

$$g_{\mu v} = \int (\partial p(X|Q)/\partial Q\mu)(\partial p(x|Q)/\partial Qv).p(X|Q)d\{x_i\}.$$

The introduced Fisher metric is a Riemannian metric. Thus we can define distance among states as well as other invariant functional such as affine connection $\Gamma^{\sigma}_{\lambda v}$, curvature tensor $R^{\lambda}_{\mu v k}$, Ricci tensor $R^{\mu k}$, and curvature scalar R *(12)* which describes the statistical manifold.

The importance of studying statistical structures as geometric structures lies in the fact that geometric structures are invariant under coordinate transforms. These transforms can be interpreted as modifications of $\{x_i\}$ set by artificial tissue with different characteristics. Thus the problem of a system survival under transition from one biophysical medium to another can be formulated geometrically.

3. Dynamic Core as a Function

After mathematical formalism is defined, we are ready to introduce the concept of dynamic core. *Dynamic core* is a dynamic system that consists of a set of dynamic elements (*see* **Fig. 2**) causally interacting with each other and the environment in a way that leads to a high level of information integration within the system and emergence of hierarchical causal interactions within the system. A higher level of causal power among system elements is compared to causal interactions with an environment. *Dynamic core* can be seen as a functional cluster characterized by strong mutual interaction among a set of subgroups over a period of time. It is essential that this functional cluster be highly differentiated.

Dynamic core will be used to describe any dynamical system that has subparts acting in causal relations with each other. To measure *causal relation* some metrics are considered such as *information integration* and *causal power*. *Dynamic core* is defined as a subsystem of any physical environment that has internal *information integration* and *causal power (10, 11)* much higher than mutual *information integration* between the system and the environment.

To describe specialized subnetworks relevant to emergence of specific high-level functions, we employ the concept of functional cluster *(10, 11)*: If there are any causal interactions within the

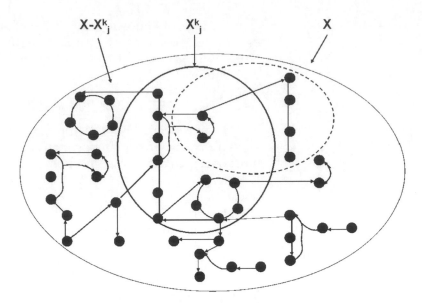

Fig. 2. System X partitioned to subset of elements X_j^k and the rest of the system $X - X_j^k$. The *dashed ellipse* represents another possible partition. The dependence between the subset X_j^k of k elements and the rest of the system $X - X_j^k{}_i$ can be expressed in terms of mutual information: $MI(X_j^k, X - X_j^k) = H(X_j^k) + H(X - X_j^k) - H(X)$.

system such as signal transfer, chain of reactions, or other related processes, the number of states that the system can take will be less than the number of states that its separate elements can take. Some subnets can strongly interact within itself and much less with other regions of the network. Geometrically, it is equivalent to higher positive curvature R ~ CI $(X_j^k) = I(X_j^k)/$ MI $(X_j^k;$ $X - X_j^k$ in the area of information manifold which reflects the system X_j^k state dynamics due to the loss of information entropy "H." The loss is due to interactions among the system elements $- I(X_j^k) = \Sigma H(x_i) - H(X_j^k)$ and interaction with the rest of the system described by mutual entropy MI $(X_j^k; X - X_j^k)$.

Curvature R near 1 indicates subnets which are as interactive with the rest of the system as they are within their subset. On the other hand, an R much higher than 1 will indicate the presence of a *functional cluster* – a subset of elements which are strongly interactive among themselves but only weakly interactive with the rest of the system. Localized function cluster is a manifestation of specialized region involved in the generation of high-level function. Geometrically, it will emerge as a horn area on information manifold.

Evolution of causal interactions among functional clusters is described by Ricci tensor R_{kj} which is a geometric analog to the concepts of *effective information* and *information integration (10)*. Effective information EI $(X_j^k \rightarrow X - X_j^k)$ between subnet X_j^k and $X - X_j^k$ can be seen as an amount of informational entropy that $X - X_j^k$ shares with X_j^k due to causal effects of X_j^k on $X - X_j^k$.

Holographic representation is spatially distributed and as such will be manifested by a smooth curvature across the manifold with fewer dimensions.

4. Evolution of the Statistical Manifold (AdS Formalism)

Evolution of the network system can be modeled by Euler–Lagrange equations taken from small virtual fluctuation of metric for scalar invariants. One example of such evolution equation is

$$J = -1 / 16\pi \int \sqrt{\det g^{\mu k}(Q)} R(Q) dQ.$$

Real-world open system such as a neural network usually requires external constraints. Those constraints represent mass transfer, mass and energy cycling, and conservation during chemical reactions as well as entropy flow for dissipative systems *(12)*:

$dS/dV \, dt = \text{const} \rightarrow \min$ – "minimum value of the entropy production by the "V" compartment"

$dG = dH - T \, dS$ – "free energy cycling".

External constraints can be added as a scalar term dependent on arbitrary covariant tensor $T^{\mu k}$:

$$J = -1 / 16\pi \int \sqrt{g(Q)} R(Q) dQ + 1 / 2 \int \sqrt{g(Q)} T^{\mu k} g_{\mu k} \, dQ.$$

That leads to the well-known geometrical equation $R^{\mu k}(Q) - g^{\mu k}(Q) R(Q) + 8\pi T^{\mu k}(Q) = 0$ which describes metric evolution. Solutions of this equation represent statistical systems under certain constraints defined by tensor $T^{\mu k}$. Thus, functionality of network systems can be presented in geometric terms where AdS *(6)* model is one solution.

5. CFT Formalism

Another way to employ geometric approach is definition of a tangent vector space "TM(X)" for each point X of manifold $M(X)$ as: $A_\mu \sim \partial \ln(p(x))/\partial x_\mu$ where Lie brackets $[A_\mu A_\nu] \sim A_k$, give us a way to find transformations which will provide invariant descriptors. First we can employ the approach developed in gauge theories that are usually discussed in the language of differential geometry that make it plausible to apply for informational geometry. Mathematically, a *gauge* is just a choice of a (local) section of some principal bundle. A gauge transformation is just a transformation between two such sections. Note that although gauge theory is mainly studied by physics, the idea of a connection is not essential or central to gauge theory in general. We can define a gauge connection on the principal bundle. If we choose a local basis of sections, then we can represent covariant derivative by the connection form A_μ, a Lie-algebra valued 1-form which is called the gauge potential in physics. From this connection form we can construct the curvature form F, a Lie-algebra valued 2-form which is an intrinsic quantity, by

$$F_{\mu\nu} = \partial_\mu A_\nu - \partial_\nu A_\mu - ig\left[A_\mu A_\nu\right].$$

$$\left[D_\mu D_\nu\right] = -ig F_{\mu\nu},$$

where $D_\mu = \partial_\mu - ig A_\mu$; $A_\mu = A_\mu^a t^a$ and t is generator of infinitesimal transformation.

Thus we can write the Lagrangian: $1/4 \ F^a_{mn} \ F^a_{mn} \square -1/2 \mathrm{Sp}$ $(F_{mn}F_{mn})$ which is invariant under transformation of coordinates. Such approach provides us with an analog of CFT Yang–Mills model on our statistical manifold.

At the same time, another evolution functional $J = \int(R + |\nabla f|^2)\exp(-f)\mathrm{d}V = \int(R + |\nabla f|^2)\mathrm{d}m$, which is dependent on function f gradient vector field defined on the manifold of volume V. It is well known in and string theory *(5)* functional that can lead us to AdS model. Thus using statistical geometry approach, we can equivalently formulate any network system functionality in terms of two models: statistical analogy of super-gravity AdS and Yang–Mills.

Figure 3 illustrates correspondence between localized and holographic representations. Each point on the toy three-dimensional manifold represents a state of the system which can be equivalently represented as a concentric conic on the two-dimensional manifold.

Such dual representation gives us an insight on how specific dynamic system localized in parametric space can be reimplemented by its holographic representation. We can state that any localized dynamic system can be equivalently represented by its holographic representation and vice versa. That representation is distributed across parametric manifold of fewer dimensions but it provides identical functionality.

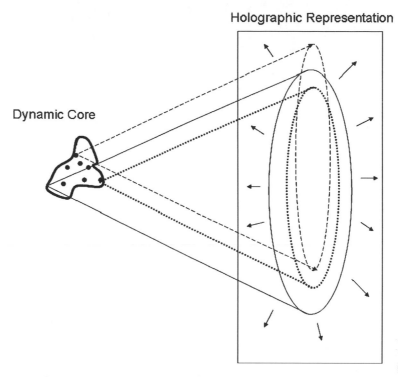

Holographic Representation

Dynamic Core

Fig. 3. Correspondence between two models. Local representation of a system on 3D statistical manifold corresponds to a set of concentric conics on 2D manifold.

6. Geometric Flow as Delocalization and Function Conservation

As we propose, an evolution of the dynamic biological system can be analyzed in terms of information geometry defined on statistical manifolds. We define two functionals to describe evolution of a statistical manifold in dual representation: AdS and CFT; it can be easily shown that one functional $J = \int(R + |\nabla f|^2)dm$ can be taken as the gradient flow $dg_{ij}/dt = 2(R_{ij} + \nabla_i \nabla_j f)$ that is generalization of the geometric flow called Ricci flow $dg_{ij}/dt = -2R_{ij}$. The interesting thing about the Ricci flow is that it can be characterized among all other evolution equations by infinitesimal behavior of the fundamental solutions of the conjugate heat equation. It is also related to the holographic renormalization group flow *(6, 7)* (RG-flow) which provides a structural form of the Ads–CFT correspondence.

Thus, transition of the system architecture can be seen as a geometric flow of information entropy on predefined statistical manifold. Results *(5–9)* demonstrate that the Ricci flow can be considered as a renormalization semigroup that distributes informational curvature over the manifold but keeps scalar invariant $R = R_{min} V^{2/3}$ which is R-curvature and V-volume of the functional cluster on information manifold. As we mentioned before, the region with strong curvature is interpreted as a subsystem with high information integration and complexity. Thus, we can see Ricci flow as a process of delocalization which migrates functional cluster to distributed representation. Based on Perelman work *(5)* for the solutions to the Ricci flow $(d/dt\, g_{ij}(t) = -2R_{ij})$ the evolution equation for the scale curvature on Riemann manifold is

$$d \,/\, dtR = \Delta R = 2\left|\text{Ric}\right|^2 = \Delta R + 2\,/\,3R^2 + 2\left|\text{Ric}^\circ\right|^2$$

It implies the estimate $R^t_{min} > -3/(2(t + 1/4))$ where the larger the t-scalar parameter is, then the larger is the distance scale and the smaller is the energy scale. The evolution equation for the volume is $d/dt\,V < R_{min}V$. Taking R and V asymptotic at large t, we have $R(t)V(t)^{-2/3} \sim -3/2$. Thus, we have Ricci flow as a process of delocalization. If V is growing, then R is decreasing and the CFT model defined on the conformal boundary of the manifold. That provides distributed representation under architecture transition **(Fig. 4)**.

In the context of artificial tissue engineering, Ricci flow provides a way to analyze the entropy flow within the system which dissolves functional clusters but preserves the system functions. These functions are distributed over parts of the system. It also provides a constraint on adjustment process for consistency between original biochemical networks and artificial tissue.

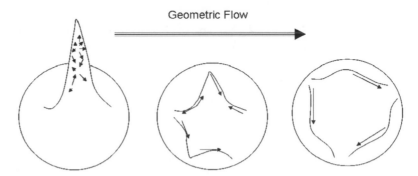

Fig. 4. Delocalization of functional cluster which is represented by R-curvature. Evolution of the manifold presented as a geometric flow.

7. AutoPoetic Functionality as Coherent Structure

Now, we are looking for all functional structures which can be reimplemented during modification in the system architecture.

As we mentioned before, deformations of the statistical manifold represent adjustment processes which might take place within the tissue. Those processes should preserve vital functions of the biological system. In terms of geometry, vital functions should be invariant under the manifold deformations. To discover invariant structures, we investigate topological evolution of information manifold.

A function is defined by an open set of causally interacting elements which is an auto-poetic *dynamic core* of the function. Thus *dynamic core* is defined geometrically as a coherent structure such as deformable connection domain on information manifolds with certain topological properties. We consider topological properties of such space that stay the same under continuous transformations. For the scalar function R evolves in time with some velocity VR; following the method of Cartan *(9)*, a certain amount of topological information can be obtained by the construction of the Pfaff sequences based on the 1-form of Action, $A = ds = VR_\mu(x)dx^\mu$ a differential constructed from the unit tangent information integration velocity field VR. We claim that emergent states are coherent topological structures:

Topological Action (energy): A

Topological Vorticity (rotation): $F = dA$

Topological Torsion (entropy): $H = A \wedge dA$

Topological Parity (dynamic core): $K = dA \wedge dA$

The rank of largest nonzero element of the above sequence gives Pfaff dimension of an information manifold. It gives us the minimal number M of functions required to determine the topological properties of the given form in a pregeometric variety of dimensions N. We require at least dimension 4 to accommodate complex systems with *dynamic cores*. This Pfaff dimension is an invariant of a continuous deformation of the domain and thus it is invariant under geometric flow.

It may be demonstrated from de Rham's theorem and Brouwer's theorem *(9)* that the odd-dimensional set (1, 3, 5,...) may undergo topological evolution but even dimensions (2, 4,...) remain invariant. It implies that coherent topological structure once established through evolution of the Pfaff dimension from 3 to 4 then will remain invariant. Thus we have another constraint that *dynamic cores* will survive any changes in biological network if those changes do not alter even Pfaff dimension of the network information manifold.

8. Notes

We propose the information geometry on statistical manifolds to analyze the evolutionary processes in biological networks. Causal relations among system elements were embodied as geometric structures on a statistical (information) manifold.

It was demonstrated that arbitrary network functionality can be represented by mathematical formalism used in string theory which is employed on statistical manifolds. Also, the same functionality can be presented in terms of conformal gauge field theory (CFT) defined on the conformal boundary of the manifold. $d + 1$-dimensional AdS model was assigned to the statistical manifold which represents a biological network. At the same time, d-dimensional CFT model was suggested to be a holographic representation of the same network.

Also, we demonstrate precise correspondence between two formalisms. And Holographic Renormalization Group Flow is suggested as a mechanism of transition from one representation to another. Such transition can conserve topological properties and preserve causal relations within the whole system.

Recovery constraint was formulated such as: for each biological network (represented by $d + 1$-dimensional compact manifold), we can construct distributed representation of the same system on d-dimensional conformal boundary.

If this strategy was implemented by biological neural networks, that might explain the recovery of some functions which were lost during stroke or cancer. The Ricci flow on statistical manifolds is suggested as a formal process for recovery of the lost functions. Based on our formal model, we suggest that for a therapy which is able to effectively implement Ricci flow as an entropy flow within the system, that system will successfully adjust the biological network and recover lost functions.

Finally, we interpreted holographic representation as a chemical network provided by the system environment. And we suggest a hypothesis *that any causal relation within a biological network has its analog in the causally related chemical reactions provided by the system environment.* Thus any modifications in the system should be consistently reflected by an environment and vice versa. Developed formalism provides a way to estimate such consistency.

References

1. Bernhard O. Palsson, *System Biology*. Cambridge University Press, Cambridge
2. Andrei L. Osterman, Tadhg P. Begley, *A Subsystems-Based Approach to the Identification of Drug Targets in Bacterial Pathogens*. Progress in Drug Research, Vol. 64
3. Anti De Sitter Space and Holography. Edward Witten (http://arxiv.org/abs/hep-th/9802150)

4. http://en.wikipedia.org/wiki/Information_geometry,http://en.wikipedia.org/wiki/Curvature_tensor

5. Grisha Perelman (November 11, 2002) "The Entropy Formula for the Ricci Flow and its Geometric Applications". arXiv:math.DG/0211159

6. Pedro Lauridsen, Ribeiro Renormalization Group Flow in Algebraic Holography http://www.arxiv.org/abs/hep-th/0306024

7. Sebastian de Haro, Kostas Skenderis, Sergey N. Solodukhin, Holographic Reconstruction of Spacetime and Renormalization in the AdS/CFT Correspondence http://www.arxiv.org/abs/hep-th/0002230

8. Sergey N. Solodukhin, Entanglement entropy and the Ricci flow http://www.arxiv.org/abs/hep-th/0609045

9. Robert M. Kiehn, (1990) "Topological Torsion, Pfaff Dimension and Coherent Structures", in: H. K. Moffatt and T. S. Tsinober eds., Topological Fluid Mechanics. Cambridge University Press, Cambridge, 449–458 http://www22.pair.com/csdc/pd2/pd2fre1.htm

10. Gerald Edelman (2006) Theories and Measures of Consciousness: An Extended Framework. 10.1073/pnas.0604347103

11. Giulio Tononi, Olaf Sporns (2003) Measuring Information Integration., *BMC Neuroscience*, **4**:31

12. Ilya Prigogine, Dilip Kondepudi (1998) *Modern Thermodynamics: From Heat Engines to Dissipative Structures.* Wiley, Chichester

Chapter 7

Protein Structure Prediction Based on Sequence Similarity

Lukasz Jaroszewski

Summary

The observation that similar protein sequences fold into similar three-dimensional structures provides a basis for the methods which predict structural features of a novel protein based on the similarity between its sequence and sequences of known protein structures. Similarity over entire sequence or large sequence fragment(s) enables prediction and modeling of entire structural domains while statistics derived from distributions of local features of known protein structures make it possible to predict such features in proteins with unknown structures. The accuracy of models of protein structures is sufficient for many practical purposes such as analysis of point mutation effects, enzymatic reactions, interaction interfaces of protein complexes, and active sites. Protein models are also used for phasing of crystallographic data and, in some cases, for drug design. By using models one can avoid the costly and time-consuming process of experimental structure determination. The purpose of this chapter is to give a practical review of the most popular protein structure prediction methods based on sequence similarity and to outline a practical approach to protein structure prediction. While the main focus of this chapter is on template-based protein structure prediction, it also provides references to other methods and programs which play an important role in protein structure prediction.

Key words: Protein structure prediction, Sequence homology, Protein sequence alignment, Comparative modeling, Fold recognition

1. Introduction

Experimental determination of protein structure remains a time-consuming and costly process and is often unsuccessful. Because of tremendous progress in genome sequencing technology the number of experimentally determined protein sequences is growing at an exponential pace. The pace of experimental determination of protein structures is also increasing but it remains

Vadim Astakhov (ed.), *Biomedical Informatics,* Methods in Molecular Biology, vol. 569
DOI 10.1007/978-1-59745-524-4_7, © Humana Press, a part of Springer Science+Business Media, LLC 2009

much slower. Protein structure prediction methods are filling the gap between the rapidly growing number of known protein sequences and known protein structures.

The entire field of homology-based protein structure prediction is based on the fact that similar sequences fold into similar structures *(1)*. It means that, when the sequence of interest or its part is globally similar to a sequence of known protein structure, then it is possible to align their residues and build a three-dimensional model of the protein using known structure as a template *(2–4)*. This method, called comparative modeling (or homology modeling), is widely believed to be the most reliable way of predicting protein structure from its sequence. In the first step, sequence–template alignment is provided as input to modeling program which then replaces residues of the template according to this alignment, builds structure fragments which are not aligned with the template, and often also refines the overall geometry of the model. The accuracy of the final model depends primarily on the (unknown) similarity between the structure which is being modeled and the structure used as a template and on the accuracy of the alignment.

Quite often, some regions of the protein sequence are not similar to known structures and then, in order to derive some structural information about them, one has to combine the comparative modeling method with methods of predicting local structural features. The methods for predicting local structural features are not the main focus of this chapter but it provides a list of selected, publicly available methods, since they are complementary to template-based predictions.

Several stand-alone programs and web servers for comparative modeling and for predicting different local structural features are publicly available. There is a growing need for automated services integrating results from different servers and for methods providing consensus predictions. Several groups are developing protein structure prediction meta-servers and integrated prediction services, which serve this purpose.

The next sections provide an overview of protein structure prediction and describe methods of detecting sequence similarity, model building algorithms, and model evaluation programs. They are followed by a short overview of automated structure prediction services and a step-by-step guide to template-based protein structure prediction.

A growing number of bioinformatics algorithms are becoming available as web servers. Quite often it is possible to obtain a useful protein structure prediction including a three-dimensional model just by using public web servers. The information about such servers is given in **Subheading 6**.

2. Overview of Protein Structure Prediction

The process of protein structure prediction is usually described as consisting of three major stages:

- Detection of sequence similarity to known structures
- Model building
- Model evaluation and improvement

In reality, the procedure is usually more complex (*see* **Fig. 1**). The first task, which needs to be performed, is the determination of domain architecture of a protein of interest. This step is often trivial for one-domain proteins and for proteins which can be fully aligned with known structures, but for long eukaryotic proteins, where only fragments can be reliably aligned with structural templates,

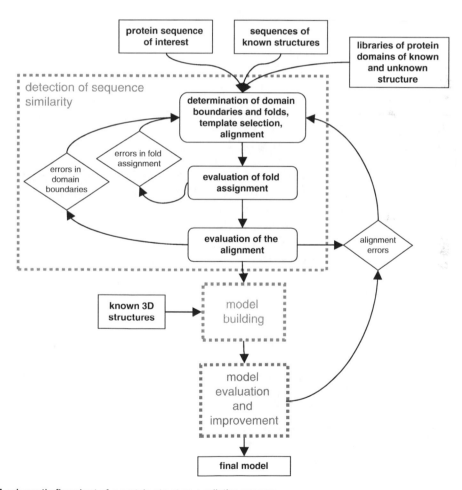

Fig. 1. A schematic flowchart of a protein structure prediction process.

it may be the most challenging part of the modeling process. Once domain architecture of a protein is determined, one can focus on choosing optimal modeling templates for individual domains and optimize the alignment between the sequence and modeling template(s). Sequence–template alignment is the most important input information for the next step when the actual model is built.

In straightforward cases of protein structure prediction these steps can be performed in sequential and automated manner. In more challenging situations errors introduced at a certain step can often be detected only by performing the next step. In such situations one can go back to the previous step and make necessary corrections. For example, errors in the prediction of domain boundaries may become visible only when alignments of individual domains are evaluated. In a similar way, errors in sequence–template alignment often become apparent only after a three-dimensional model of the structure is built. A comprehensive review of the field of protein structure prediction has been recently published by Xiang *(5)* and most recent developments have been reviewed by Ginalski *(6)*.

3. Detecting Sequence Similarity to Known Structures

3.1. Assignment of Domain Boundaries and Folds

As the first important and often forgotten step of protein structure prediction, one should determine approximate boundaries of structural domains and, if possible, assign them to annotated protein families and folds. Prediction of domain architecture itself may be sufficient for some practical purposes such as proposing a hypothesis about protein function and assigning it to a functional category or family. Splitting a query sequence into putative domains and treating them as separate prediction problems usually improves prediction accuracy. For instance, sequence profiles obtained for individual domains are usually less prone to profile contamination caused by unrelated and misaligned sequences than profiles built for full multi-domain sequences.

Quite often, the initial sequence similarity searches performed against libraries of structural or functional protein domains are sufficient for assignment of domain boundaries and their folds. Significant similarity between a fragment of a query sequence and a known protein domain provides direct information about this domain's location and its boundaries. The remaining long fragments of the query sequence, which cannot be aligned with known domain(s), are natural candidates for other (possibly novel) structural domains. Such fragments are usually subject to separate searches with homology detection algorithms (*see* **Notes 2** and **3**).

It should be noted that, while detection of significant sequence similarity to known protein domains provides accurate information about domain boundaries, the accuracy of such prediction depends on the features of the alignment and the accuracy of the boundaries of the annotated domain used for this prediction. High scoring and complete alignment with solved protein structure or individual structural domain provides a very reliable prediction of domain boundaries and, if such prediction is available, the problem of determining domain boundaries is solved (*see* **Note 1**). However, if the alignment covers only part of the domain used for the prediction, then determination of exact domain boundaries becomes more difficult (*see* **Note 5**).

Several libraries of annotated protein domains, such as SCOP *(7)*, PFAM *(8)*, PRODOM *(9)*, and INTERPRO *(10)*, are publicly available (*see* **Table 1**), and, with the exception of SCOP, only about half of domains included in these libraries have structural templates. If some protein fragment cannot be matched with any structural domain, then even domains which are not based on structural templates can serve as some approximation of structural domains. However, one should bear in mind that such domains do not always correspond to structural domains. They are usually results of automated domain prediction and only, in some cases, are curated by experts in a given protein family.

For fragments with no detectable similarities to annotated protein domains or known structures, the information about possible domain boundaries can be derived from multiple alignments of all homologous sequences. If some group of homologous sequences forms "blocks" aligned only with some region of the query, then such region is likely to form a separate structural domain.

Table 1
Selected publicly available databases of protein domains

Method/server	Method of determining domain boundaries	URL and reference
SCOP and SUPER-FAMILY	Based on structures, expert curated	http://scop.mrc-lmb.cam.ac.uk/scop/ *(7)*
		http://supfam.org/SUPERFAMILY/ *(11)*
PFAM	Based on structures and multiple sequence alignments, expert curated	http://pfam.janelia.org/ *(8)*
PRODOM	Based on the analysis of multiple sequence alignments. Takes into account information from SCOP database of structural domains	http://prodom.prabi.fr/ *(9)*
InterPro	Integrates protein domain annotations from several databases	http://www.ebi.ac.uk/interpro/ *(10)*

Table 2
Selected algorithms for protein domain prediction

Method/server	Methodology	URL and reference
DOMPro	Profiles, secondary structure, relative solvent accessibility, and recursive neural networks	http://www.ics.uci.edu/~baldig/domain.html *(12)*
DOMAC	Template-based and ab initio methods	http://www.bioinfotool.org/domac.html *(13)*
GlobPlot	Amino-acid propensities	http://globplot.embl.de/ *(14)*
DomPred	Template-based prediction, secondary structure element alignment	http://bioinf.cs.ucl.ac.uk/dom-pred/ *(15)*
CHOP	Template-based prediction	http://cubic.bioc.columbia.edu/services/CHOP/ *(16)*

Several specialized algorithms for protein domain prediction are publicly available (*see* **Table 2**). Some of them utilize the template-based prediction method described above but many of them also provide ab initio predictions of domain boundaries when structural templates are not available.

3.2. Methods for Detecting Sequence Similarity

The methods of detecting sequence similarity provide necessary input information for model building algorithms. They identify modeling templates, provide sequence–template alignment, and play an important role in the determination of domain boundaries. The crucial issues of detection sensitivity and alignment accuracy are briefly discussed below. A comprehensive review of sequence comparison methods has been recently published by Dunbrack *(17)*.

3.2.1. Sensitivity of Fold Prediction

The structure of a novel protein can be modeled when it is possible to recognize its similarity to a protein with known structure. However, standard sequence comparison methods lose sensitivity in the so-called "twilight zone" where sequence identity between proteins falls below 30% *(18)*. The sensitivity of fold recognition and accuracy of the corresponding sequence–structure alignment can be improved by using evolutionary information extracted from large sets of related protein sequences. In brief, such methods put higher weights on residues conserved in a protein family and, thus, more likely to be conserved in its remote homologues. This assumption led to the development of a group of methods where, instead of aligning two protein sequences, one aligns a protein sequence with an entire protein family. The protein family can be represented by a sequence profile as it is implemented in PSI-BLAST *(19)*, or by Hidden Markov Model (HMM) *(20)*. A natural next step in this strategy is to align two sequence profiles, as it is implemented in FFAS *(21)* (and several other methods) or two Hidden Markov Models, as implemented in HHSEARCH *(22)*.

The application of sequence profiles significantly increases the number of reliable fold predictions which can be made. In order to evaluate homology prediction methods one can test them on representative sets of known structures and calculate the percentage of correct predictions at different error levels. ASTRAL resource *(23)* of the SCOP database *(7)* has been used to construct such a benchmark consisting of 5,868 structures of protein domains sharing less than 25% sequence identity. BLAST, PSI-BLAST, and FFAS were used as representatives of sequence–sequence, profile–sequence, and profile–profile comparison methods. The results clearly illustrate advantages of techniques based on sequence profiles (*see* **Fig. 2**). With 5% of false positives, profile–sequence comparison method PSI-BLAST *(19)* yields two times more correct predictions than the sequence–sequence comparison method BLAST *(24)*. FFAS, which is based on the profile–profile alignment, improves the sensitivity by another 20%.

Other advanced fold recognition methods which use evolutionary information include 3D-PSSM *(25)*, FUGUE *(26)*, BIOINBGU *(27)*, PROSPECT *(28)*, and SAMT98 *(29)*. These methods are more sensitive than sequence–sequence and profile–sequence alignment methods.

3.2.2. Alignment Accuracy

In general, alignment accuracy is related to the alignment score and sequence identity. For every alignment method, better similarity scores and higher sequence identity indicate that structural

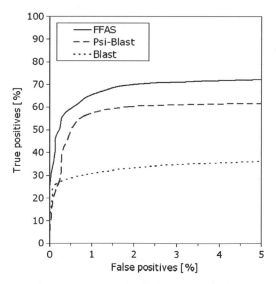

Fig. 2. The percent of correct and incorrect structural predictions obtained with sequence–sequence, profile–sequence, and profile–profile alignment methods (represented by BLAST, PSI BLAST, and FFAS, respectively). The test has been performed on the representative benchmark set of pairs of homologous protein domains selected from the SCOP database. Protein pairs with sequence identity exceeding 25% have been removed from this set.

templates are more similar to the query sequence and the accuracy of the alignment is higher (*see* **Note 6**). At the same time, average lengths of alignments are different in different methods, making the comparison difficult. The alignments obtained by profile–profile and other advanced comparison algorithms are usually longer than alignments from profile–sequence or sequence–sequence comparison methods. The elongation of the alignment by these methods is, in most cases, justified, adding fragments of similar structure. The accuracy of the alignment produced by different methods can be compared using benchmarks similar to those used to assess sensitivity of fold recognition. The alignment calculated with different methods can then be compared to structural alignments which are often used as "the standard of truth." Another way of comparing alignments is to use them to build a simple model of one protein using the other as the template and then evaluate similarity between the model and the real structure by standard measures of structural similarity such as CαRMSD. By using such benchmarks it has been demonstrated that profile–sequence comparison methods produce more accurate alignments than traditional sequence–sequence comparison methods, and profile–profile matching algorithms give alignments which are even more accurate than profile–sequence alignments *(30)* (*see* **Table 3**). Several advanced alignment and homology recognition servers are publicly available (*see* **Table 4**)

3.3. Methods of Predicting Local Structural Features

The methods of predicting local structural features are usually based on the comparison of local features of protein sequence with the statistics derived from known protein structures. This general approach, implemented in various mathematical and algorithmic procedures, is used by the methods of predicting secondary structure, structural disorder, coiled-coil regions, transmembrane helices, and signal peptides. It means that, even when the protein sequence of interest is not globally similar to any sequence of known structure, it is possible to predict, with cer-

Table 3
The comparison of sequence–sequence, profile–sequence, and profile–profile alignments

Alignment method	Agreement with structural alignment%	CaRMSD (Å)
Sequence–sequence	39	7.8
Profile–sequence	47	5.3
Profile–profile	48	4.4

Table 4
Selected methods for detecting sequence similarity and sequence alignment

Method/server	Methodology	URL and reference
BLAST	Sequence–sequence	http://www.ncbi.nlm.nih.gov/BLAST/ (24)
PSI-BLAST	Profile–sequence	http://www.ncbi.nlm.nih.gov/BLAST/ (19)
HMMER	Sequence–HMM	http://hmmer.janelia.org/ (31)
SAM	Sequence–HMM	http://www.soe.ucsc.edu/research/compbio/sam.html (29)
FUGUE	Sequence–profile	http://www-cryst.bioc.cam.ac.uk/fugue/ (26)
Phyre	Profile–profile	http://www.sbg.bio.ic.ac.uk/phyre/ (25)
FFAS	Profile–profile	http://ffas.burnham.org/ (21)
HHPred	HMM–HMM	http://toolkit.tuebingen.mpg.de/hhpred (22)

Table 5
Selected methods for predicting local structural features

Program/ server	Predicted/detected feature	URL and reference
TMHMM	Transmembrane helices	http://www.cbs.dtu.dk/services/TMHMM-2.0/ (32)
COILS	Coiled-coil regions	http://www.ch.embnet.org/software/COILS_form. html (33)
DISOPRED	Structural disorder	http://bioinf.cs.ucl.ac.uk/disopred/ (34)
PSIPRED	Secondary structure	http://bioinf.cs.ucl.ac.uk/psipred/ (35)
SEG	Low-complexity regions	ftp://ftp.ncbi.nih.gov/pub/seg/seg/ (36)

tain accuracy, which fragments of the sequence will form alpha-helices, beta-strands, or loops, and which are unlikely to form any well-defined structure. It is also possible to predict where coiled-coil regions and transmembrane helices will be formed. The review of methods of predicting local protein features is far beyond the scope of this chapter. **Table 5** lists selected methods of predicting local protein structure, which can complement template-based predictions in regions without detectable similarity to known protein structures.

3.4. Evaluation and Improvement of Fold Assignment

Homology recognition methods do not always provide a convincing prediction and, sometimes, give incorrect predictions. Even profile–profiles methods can detect only 38% of structural similarities between proteins from different SCOP families at the error level of 3%. It means that more than half of nontrivial structural similarities between proteins can only be recognized, when their structures are known. However, even when direct searches do not give compelling results, it is often possible to obtain useful prediction or to evaluate weak prediction by using special strategies and criteria. They include evaluating the consistency of the prediction, performing intermediate sequence similarity searches, and evaluating the quality of sequence profiles.

3.4.1. The Consistency of the Prediction

By checking the consistency of fold assignment one can get additional information about its reliability. For example, results of sequence similarity searches against the SCOP database usually contain structure classification that enables an easy check of structural consistency of the prediction. In particular, if the same region of the query sequence is aligned with two SCOP domains of completely different folds, then one or both of these predictions are incorrect. On the other hand, if a weak prediction with a very marginal score contains hits to different proteins of the same fold but significantly different sequences, then its reliability may be interpreted as higher than it would be if based solely on the score (*see* **Note 4**). The criterion of structural consistency is an integral part of some prediction methods such as 3D-JURY *(37)*.

3.4.2. Intermediate Sequence Similarity Searches

The accuracy of the prediction depends on the features of sequence profiles used for this prediction, which in turn may depend on the specific query used to "seed" database searches which produce the profile. Thus, performing multiple sequence similarity searches started with homologues of the original query sequence may be helpful in difficult prediction problems. The results of searches obtained for sequences homologous to the query can be more (or less) convincing than the results obtained with the original query. The predictions obtained with homologous sequences should be consistent, as described above. However, one should bear in mind, that the results of searches involving intermediate steps of low reliability are significantly less reliable than results of direct searches, since the probabilities of errors accumulate.

3.4.3. The Quality of Sequence Profiles

Since the most effective methods of detecting remote homologies are based on the comparison of sequence profiles, their effectiveness critically depends on the quality of these profiles. For example, it has been shown that FFAS predictions obtained with profiles which include larger numbers of homologues are more accurate than predictions based on the alignment of profiles which included smaller numbers of homologues *(21)*. It is always informative to examine multiple sequence alignments used to derive profiles. In the case of FFAS these alignments can be displayed directly by the server. The rule of thumb is that the most reliable profiles include large numbers of sequences which share the same well-conserved regions. Profiles based on the alignments containing large numbers of low-complexity or coiled-coil proteins and lacking well-conserved sequence motifs often yield incorrect predictions. Because of their highly biased compositions and repetitive character, sequence similarities in such regions cannot be reliably detected. They are often incorrectly recognized as similar to unrelated protein sequences, causing so-called "profile contamination." In general, low-complexity or coiled-coil regions should be removed from the query sequence. In many profile-based methods such regions are also masked in the sequence database used to build profiles. Low-complexity regions of protein sequences can be detected with the SEG program *(36)* and coiled-coil regions can be predicted with the COILS method *(33)*.

3.5. Evaluation and Improvement of Sequence Alignment

Determination of the alignment between the sequence of interest and the modeling template is a decisive step in comparative modeling. The quality of the alignment determines the quality of the final model *(38)*. This is especially the case for pairs of remote homologues which can be reliably recognized only by fold recognition methods. Such algorithms can recognize very distant homologues but alignments of such pairs may be often incomplete or partly incorrect. However, the fact that remote sequence similarity has been detected by sequence comparison means that some part of the alignment produced by that method is accurate. Usually, such parts correspond to the most conserved regions of the protein family. In the pairwise alignment they usually correspond to regions with the highest density of conserved residues (*see* **Fig. 3**). It is often (but not always) located in the hydrophobic core of the structure or in the functionally important regions such as active site or binding site. Unreliable regions of the alignment are fragments flanking gaps, terminal parts of the alignment, and fragments of low sequence similarity between the model and the template (*see* **Fig. 3**). In the template structure they correspond to loops or fragments of structural disorder.

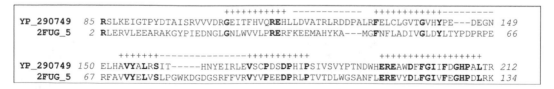

Fig. 3. Profile–profile alignment calculated with FFAS used for modeling of NADH dehydrogenase subunit C from *T. fusca* (Genebank id YP_290749). The only structure homologous to this protein was subunit 5 of respiratory complex from *T. thermophilus* (PDB code 2fug chain 5). FFAS aligned 66% of the sequence of YP_290749 with the sequence of chain 5 of 2fug, with a score of −79 and a sequence identity of 27%. Regions of low and high alignment reliability are marked with *minuses* and *pluses*, respectively. Since only one template structure was available, inaccurate regions of the model could not be identified from structural superposition of homologous structures.

4. Model Building

4.1. Methods for Model Building

In the final step of protein structure prediction the actual three-dimensional model of a protein of interest is built. The process of model building can be roughly divided into three tasks:

- Replacing template residues with query residues according to the alignment and predicting conformations of side-chains

- Building regions of protein structure which are not aligned with the template

- Model refinement

While useful for illustration purposes, this description is a simplification. In reality, different modeling algorithms may perform only some of those tasks and in other methods some tasks may be integrated in to one step.

The following sections describe key elements of the model building process and present popular, publicly available methods. An excellent comprehensive comparison of the most popular model building algorithms has been published by Wallner and Elofsson *(39)*.

4.1.1. Prediction of Side-Chain Conformations

The first task is often referred to as "side-chain packing." It can be described as combinatorial search for optimal packing of new side-chains placed on the template's backbone. It is known to be the most accurate and reliable part of the model building process, if the differences in backbone geometry between the template and the structure which is being built are small. The accuracy of side-chain packing prediction is expected to be much higher in the protein core, since it is more densely packed and the geometry of the backbone is much better conserved. Some algorithms perform only this step of model building and, in fact, such simple models may be sufficient for many purposes. For example, one can be interested only in modeling of the active site which is often completely aligned with the modeling template. Moreover, for some purposes building additional elements of the structure may be less important than ensuring maximal accuracy of the model (for example, this is often the case for models used for molecular replacement phasing method of X-ray crystallography).

4.1.2. Prediction of Unaligned Regions

Short fragments of the query sequence which are not covered by the alignment can be built using loop prediction methods. These methods usually rely on large libraries of loop conformations derived from known protein structures. It is widely accepted that this procedure has good accuracy for sequence fragments shorter than five or six residues, however, accurate predictions have been reported for much longer fragments *(40, 41)*. Loop prediction algorithms can be divided into two groups: database-based methods and ab initio prediction methods. Database-based methods use libraries of structure fragments extracted from PDB as putative loop conformations, test their geometric feasibility, and evaluate them with energy-like force field, while ab initio methods build loop geometry "from scratch" using only general constraints of protein backbone geometry and then evaluate them using energy force field. The most accurate and efficient loop prediction algorithms became part of modeling programs such as Modeller *(42)* and Jackal *(5)*, so, for practical purposes, it is usually not necessary to use separate loop prediction programs. However, it is important to know that the accuracy of fragments of the model which were aligned to the template is usually much higher than the accuracy of fragments corresponding to gaps in the alignment. The accuracy of the latter strongly depends on their length (*see* **Note** 7).

4.1.3. Model Refinement

Some modeling methods perform global optimization of the model's geometry using potentials or probability distributions based on statistics derived from known protein structures. This procedure can correct significant errors in the model's geometry in the regions where it violates regions of local conformational space never occupied by protein structures. On the other hand, global optimization may alter conformations of regions which were already highly conserved between the query and the template, resulting in local decrease of the model's accuracy. Thus, if high accuracy of the conserved protein core is most important, then one may consider using modeling methods which do not refine global geometry of the model.

4.1.4. Available Software for Model Building

Several software packages for model building have been developed and most of them are freely available for academic use or even available as web servers. Recently, a comprehensive comparison of those programs has been published *(39)*. The most popular, publicly available, model building programs are listed in **Table 6**. It is important to note that most modeling programs require exact correspondence between template sequence used in the alignment and atomic coordinates of the template (for more information *see* **Note 9**).

5. Evaluation and Improvement of the Model

5.1. Assessment Based on the Comparison of Multiple Models or Templates

The accuracy of protein models depends mostly on the (unknown) similarity between the structure of interest and the modeling template. Since the structure, which is being modeled, is not known, structural similarity has to be estimated based on the similarity of their sequences. The threshold of 30–35% of sequence identity

Table 6
Selected model building programs

Method/server	Performed tasks	URL and reference
Modeller	Side-chain prediction, loop building, refinement	http://www.salilab.org/modeller/ *(42)*
SCWRL	Side-chain prediction	http://dunbrack.fccc.edu/SCWRL3.php *(43)*
What If	Side-chain prediction	http://swift.cmbi.ru.nl/whatif/ *(44)*
Jackal	Alignment tuning, side-chain prediction, loop building, refinement	http://wiki.c2b2.columbia.edu/honiglab_public/index.php/Software:Jackal *(5)*
SWISS-MODEL	Side-chain prediction, loop building	http://swissmodel.expasy.org//SWISS-MODEL.html *(45)*

is widely accepted as the limit of accurate comparative modeling, but, in reality, that relationship varies significantly between protein families, as it becomes apparent when structural alignments of large families are calculated and analyzed *(46)*. The limit of accurate modeling is thus different for different protein families. The accuracy of the model can be assessed based on the structural variability observed in known structures from a protein family. If modeling templates found by homology detection algorithm show only small structural differences, then new structures from this family are also likely to be well conserved. Models of unknown structures from such protein families could be very accurate, even when sequence identity to the closest known structure is very low. In order to assess structural similarity between known structures from a family one can use multiple structural alignment algorithms such as POSA *(47)*. The POSA server provides a quantitative measure of structural similarities between submitted structures along with a graphical interface, which is very helpful in determining the level of structural conservation in the family (*see* **Fig. 4**, **Notes 1** and **4**).

Structural alignments of protein families confirm the well-known fact that the conservation of the protein structure varies between different regions of the structure. Protein core is almost always more conserved than external parts of the structure such as loops and turns, but structure conservation may also differ between individual loops and secondary structure elements. The structural alignment of a family makes it possible to assess the level of structural conservation in different regions of the structure. Since the elements of protein structure with the lowest levels of structural conservation usually also correspond to the regions

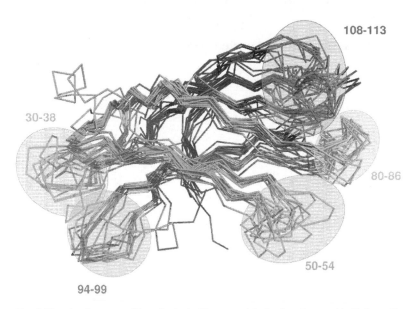

108-113

30-38

80-86

50-54

94-99

Fig. 4. Structural superposition of selected immunoglobulin domains used to find unreliable regions of the model which was based on one of them. The regions of the model, expected to contain large errors, are highlighted.

of the lowest accuracy of the alignment used for modeling, the model's accuracy in those regions is significantly lower than in highly conserved regions (*see* **Note 8**).

5.2. Assessment of the Model

Several specialized model evaluation methods have been built (*see* **Table 7**). Most of them are based on some type of energy-like function, which detects deviations from protein-like geometry and regions with a large number of unfavorable interactions between side-chains. Since the analysis of three-dimensional protein structures is much more complex than the analysis of one-dimensional protein sequences, the methods of model evaluation are less reliable than, for example, sequence comparison techniques. In general opinion they perform much better in identifying incorrect regions of the model than in global assessment of model's accuracy.

6. Meta-Servers, Integrated Prediction Services, and Other Programs Useful in Protein Structure Prediction

Automated servers for protein structure prediction vary in the extent of their functionality. Some servers such as FFAS03 *(54)* provide access to a single sequence similarity method and enable automated building of structural models. Metaservers such as PCONS *(55)*, 3D-SHOTGUN *(56)*, and 3D-JURY *(37)*

Table 7
Selected methods for evaluation of protein models

Method/server	Methodology	URL and reference
VERIFY3D	Statistical energy-like functions	http://www.doe-mbi.ucla.edu/ Services/Verify_3D/ *(48)*
ProSaII	Statistical energy-like functions	https://prosa.services.came.sbg.ac.at/ prosa.php *(49)*
PROCHECK	Stereochemical quality – covalent geometry, planarity, dihedral angles chirality, non-bonded interactions, main-chain hydrogen bonds, disulphide bonds	http://www.biochem.ucl. ac.uk/~roman/procheck/ procheck.html *(50)*
WHATCHECK	Chirality, bond lengths, bond angles, torsion angles, rings and planarity, inside/outside profile, packing quality, backbone geometry, side-chain rotamers	http://www.sander.embl-heidelberg.de/ whatcheck/ *(51)*
ANOLEA	Energy-like statistical potentials centered on heavy atoms	http://protein.bio.puc.cl/cardex/ servers/anolea/index.html *(52)*
Probe	Evaluates and visualizes inter- and intra-molecular packing	http://kinemage.biochem.duke.edu/ software/probe.php *(53)*

collect predictions from several publicly available protein prediction servers and provide a consensus prediction. This approach is widely regarded as the most effective method of predicting very remote homologies.

Table 8 contains a list of the most popular automated protein prediction services. Automated servers for protein structure prediction are regularly assessed in CAFASP *(57)* and Live-Bench *(58)* experiments. A comprehensive review and evaluation of protein prediction servers has been recently published by Fischer *(59)*.

In the most challenging problems, protein structure prediction requires a comprehensive structural and functional analysis of a protein family. If sequence identity to the closest modeling template is below 30%, and multiple modeling templates are available, then it is always useful to align and compare structures of those templates in order to study structural variability in a family and to assess the accuracy of different regions of the model. This can be done with the multiple structural alignment server POSA *(47)*. Protein structure viewers, used by POSA and by many other programs referenced in this chapter, are listed in **Table 9**.

Table 8
Selected metaservers and integrated prediction services

Method/server	Main features	URL and reference
PredictProtein	Predicts different local structural features and finds structural templates	http://www.predictprotein.org/ *(60)*
PSIPRED Server	Predicts different local structural features and finds structural templates	http://bioinf.cs.ucl.ac.uk/psipred/ *(61)*
ModWeb	Finds modeling templates and calculates three-dimensional models	http://salilab.org/modweb *(62, 63)*
BioinfoBank MetaServer	Collects predictions from different advanced homology detection tools and combines them into consensus prediction	http://meta.bioinfo.pl/ *(37)*
3D-JIGSAW	Splits the query sequence into putative domains, finds structural templates and builds 3D models	http://www.bmm.icnet.uk/ servers/3djigsaw/ *(64)*
FFAS/ProtMod	Finds structural templates using FFAS and builds models with publicly available programs	http://ffas.burnham.org/ *(54)*
XtalPred	Runs several programs predicting local protein structure and presents graphically, provides estimates probability of protein crystallization, runs PSI-BLAST against PDB database	http://ffas.burnham.org/XtalPred-cgi/ xtal.pl *(65)*
M4t	Finds structural templates using blast, uses multiple templates	http://manaslu.aecom.yu.edu/M4T/ *(66)*

7. Protein Structure Prediction Step by Step

This section describes a step-by-step procedure for protein structure prediction. The procedure is illustrated by simple examples of predictions obtained using the FFAS profile–profile alignment server.

7.1. The Objective of the Prediction

Selection of the methods used for protein structure prediction depends on the purpose of the prediction. In particular:

1. For the analysis of protein surfaces and large binding sites or for docking protein models to low-resolution maps from

Table 9
Additional programs and web servers useful in protein structure prediction

Method/server	Main features	URL and reference
Chime	Simple protein structure viewer. Works as a plug-in for Internet Explorer, it is used by structure prediction web servers such as FFAS and POSA	http://www.mdl.com/products/frame-work/chime/
Jmol	Protein structure viewer. Works as the Java applet for web browsers. Used by many structure prediction web servers and databases	http://jmol.sourceforge.net/
POSA	Protein structure comparison server. Very useful for evaluation of similarities and difference between modeling templates or models	http://fatcat.burnham.org/POSA/ *(47)*

electron microscopy, the overall shape and completeness of the model are most important. Select modeling templates covering maximal percentage of the query sequence or use multiple modeling templates. Use modeling programs such as Modeller or Nest which are capable of building fragments which are not aligned with the template.

2. For active site analysis and analysis of interactions in the protein core high precision of the most conserved regions of the structure is most important. It is recommended to select modeling templates with the highest number of conserved residues in the regions of interest. It is also often better to use modeling programs which do not alter side-chain conformations of residues which are conserved in the alignment between the template and the query sequence (for example, SCWRL or WhatIf).

7.2. Determine Domain Boundaries and Assign Folds

1. The procedure depends on the length of the query sequence. As a rule of thumb, prediction for sequences shorter than 300–500 residues can be obtained automatically using one of the automated prediction servers (*see* **Table 8**). Longer sequences can also be submitted to prediction servers but only in order to get initial predictions and to split them into putative structural domains. It is recommended to use advanced profile–sequence and profile–profile methods. It is always better to try more than one method and compare results from different methods.

2. Determine domain architecture of the query sequence and, in particular, answer the following questions: Which parts can be aligned to known structural domains or to entire structures? Which parts can be aligned only to annotated domains without structural templates? Which parts remained not aligned with any structural domain?

3. For low-scoring predictions evaluate the consistency of fold assignment. It can be done by running searches against the SCOP database (for example, using Superfamily or FFAS) or by direct structural comparison of modeling templates found by sequence similarity algorithms. At this stage one should answer the following questions: Are there any sequence fragments which are assigned to two different folds? If yes, is it possible to tell which assignment is correct?

4. Run separate sequence similarity searches against databases of known structures and known structural domains for remaining long fragments (of more than 30–50 residues) without alignment to any known structures (*see* **Notes 2** and **3**).

5. If there are still long fragments without alignment to domains of known structure, then run methods of predicting local structural features (*see* **Table 5**) for the entire sequence and for those fragments.

For examples of determining protein domain architecture *see* **Notes 1–4**.

7.3. Select Modeling Templates for Domains

1. Run separate sequence similarity searches for sequence fragments corresponding to each putative domain detected in the previous step. If domain boundaries are uncertain, run separate searches for different variants of domain boundaries (*see* **Note 5**).

2. Select optimal modeling template(s) for domains to be modeled. If there are several templates with similar scores in the fold prediction method, then select the template with highest sequence identity of the alignment (*see* **Note 6**). Note that comparing alignments by their sequence identity values makes sense only if they are of similar lengths. If there are several templates with similar sequence identity, then use the template with smaller number and length of gaps in the alignment (*see* **Note 1**). If there are no significant differences in scoring and alignment of modeling templates, then consider building multiple models and evaluating them or building a single model using multiple modeling templates.

3. If the algorithm finds templates with high sequence similarity to the query but with short alignment and templates with lower sequence similarity but longer alignment, then consider using multiple templates (this can be done, for example, with Modeller program).

7.4. Evaluate and Correct Sequence–Template Alignments

1. Identify stable and unstable regions of the alignment. Stable regions usually correspond to fragments with a small number of gaps and the highest number of conserved residues. Unstable regions are terminal regions of the alignment, short alignment segments, and regions flanking gaps, especially if these regions contain a small number of conserved residues (*see* **Fig. 3**). If alignments with different templates have different regions of high and low stability, then consider using multiple modeling templates.

2. Try to assess geometric feasibility of the alignment. In order to do that one can prepare and assess preliminary, three-dimensional models of the domain(s) of interest using fast modeling program which does not build unaligned regions (for example, SCWRL or WhatIf). In principle, if consecutive residues of the sequence which is being modeled are too distant to be connected without large conformational changes of the backbone, then it usually indicates alignment errors. Gaps inside secondary structure elements also indicate errors in the alignment. Short fragments of the model which are distant from the rest of the model are also usually incorrect (also *see* **Note 7**).

7.5. Build 3D Models

1. Build protein models using tools selected according to the purpose of the prediction. In general, it is recommended to use more than one tool and more than one modeling template. As it was mentioned, errors in template selection and alignment often become apparent only after the model is built. Make sure that residues in all models are numbered consistently with the sequence being modeled (*see* **Note 8**). Check whether orientation of domains is likely to be conserved by examining superposition of template structures (*see* **Notes 1** and **2**). If it is not the case, then bear in mind that only separate domains can be modeled.

7.6. Evaluate the Models

1. If modeling includes loop building and model refinement, then check backbone geometry with model evaluation tools (*see* **Table 7**). Regions of high deviations from ideal backbone geometry and regions with multiple unfavorable interactions probably indicate errors in the alignment which may be corrected. If it is impossible to correct them by using different alignment, consider removing them from the model (*see* **Note 7**).

2. If multiple models based on templates of *comparable similarity* to the query sequence are available, then superimpose them (for example, by using the POSA server – *see* **Notes 1** and **4**). Examine the superposition with a protein structure viewer. Regions which are well superimposed and have

consistent residue numbering also correspond to the most reliable parts of the model. Well-superimposed fragments with shifts in residue numbering between different models indicate probable errors in the alignment, which were not noticed before. Regions where each model has significantly different local structure are usually unreliable and should probably be removed from the model.

3. Introduce corrections to the alignment and, if possible, repeat the modeling step and choose the model expected to be most accurate.

8. Notes

1. *Example of straightforward protein modeling.* The example of protein TM0006 from *T. maritima* (http://www.ncbi.nlm.nih.gov/entrez/viewer.fcgi?db=protein&id=15642781) illustrates the case of straightforward domain assignment and protein modeling. The sequence of TM0006 has been submitted to the FFAS server at http://ffas.burnham.org. Two databases of profiles have been searched – PDB and SCOP. The results against SCOP indicate the presence of two different structural domains: N-terminal domain from D-glucarate dehydratase-like (TIM beta/alpha-barrel fold) and C-terminal domain from enolase N-terminal domain-like (enolase N-terminal domain-like fold). The examination of hits found by FFAS in the PDB database shows that, in structures homologous to TM0006, these two domains always coexist. In order to check whether they appear in the same orientation in homologous structures one can submit PDB codes of its structural templates to the POSA server at http://fatcat.burnham.org/POSA. The result clearly indicates that this is really the case. The top-scoring FFAS template (1jpmA) has also the highest sequence identity to the sequence of TM0006 and it is aligned with its entire sequence. Also note, that the similarity between TM0006 and 1jpmA is detectable with BLAST (see *Blast* link on the page with FFAS results against PDB). One can conclude that 1jpmA is a very good modeling template for TM0006 and one can obtain 3D model just by clicking *model* link corresponding to this template, which launches the ProtMod server which provides access to popular modeling programs. Since there are only a few one-residue gaps in the alignment, there will be no large differences between models obtained with the methods which build fragments not aligned to the template (such as MODELLER) and methods which do not (such as SCWRL).

2. *Example of domain duplication*. Protein TM0037 from *T. maritima* (http://www.ncbi.nlm.nih.gov/entrez/viewer. fcgi?db=protein&id=15642812) illustrates a more interesting example of domain assignment, where one simple search against PDB and SCOP databases does not yield all necessary information. The results of FFAS search against the PDB database indicate significant homology of the C-terminal half of the sequence to proteins from the HD-domain family (HD-domain/PDEase-like fold). The N-terminal half, however, shows only one relatively weak hit to the structure 3b57A. A separate search performed only for the first 200 residues of TM0037 against PDB and SCOP databases showed significant hits to the same domain, which was found to be similar to the C-terminal part. Clearly, TM0037 consists of two domains of the same fold but there is no such structural template in the PDB. Note that none of structural templates found for TM0037 by FFAS could have been predicted by BLAST or PSI-BLAST. In summary, it is possible to obtain fold prediction and 3D models for both domains of TM0037 but their orientation cannot be predicted, since homologous templates are available only for individual domains.

3. *Example of difficult domain assignment*. Protein TM0265 from *T. maritima* (http://www.ncbi.nlm.nih.gov/entrez/viewer. fcgi?db=protein&id=15643035) provides another nontrivial example of domain assignment. As indicated by 100% hits in preliminary FFAS and BLAST searches against PDB, database structures of N-terminal and C-terminal domains of TM0265 are solved. The central domain, however, shows only short hit to SCOP domain from the family of C-terminal UvrC-binding domain of UvrB (folding into alpha-hairpin). This assignment needed to be confirmed by a separate search performed only for the sequence of central domain and, indeed, it confirms that the central fragment of the central domain of TM0265 contains UvrC-binding domain. However, flanking regions of the central domain still did not have structural prediction and were subject to two separate searches which did not reveal any significant hits to known structures. Since structures of N-terminal and C-terminal domains are known, the real gain from protein structure modeling in the case of TM0265 is the prediction of UvrC-binding alpha-hairpin in the central part of the sequence. In order to learn more about the character of the entire central domain surrounding the predicted UvrC-binding alpha-hairpin, one can submit it to XtlaPred server and examine local structure predictions calculated with Psipred, Disopred, TMHMM, and Coils. The results indicate that the central domain of TM0265 does not contain long fragments of disordered structure and seems to have very well-defined

secondary structure with several helices, loops, and beta-strands. (The fragment which was predicted to form an alpha-hairpin, is predicted to be helical, as expected.). One can conclude that the central domain of TM0265 probably has globular character and contains UvrC-binding alpha-hairpin.

4. *Example of weak but consistent fold prediction confirmed by additional local structure prediction.* Protein TM0376 from *T. maritima* (http://www.ncbi.nlm.nih.gov/entrez/viewer.fcgi?db =protein&id=15643144) can serve as an example of weak but consistent fold prediction confirmed by the prediction of helical transmembrane domain obtained with TMHMM. The results of FFAS searches against PDB and SCOP databases contain several very weak hits to the aquaporin-like family of transmembrane proteins. The consistency of this prediction makes it much more convincing than it would be, if only one hit was detected. Moreover, the sequence has been submitted to the TMHMM server which gave very strong prediction of seven transmembrane helices. After combining this information, one can say with high confidence that TM0376 belongs to the aquaporin family and even build its three-dimensional model. The accuracy of this model can be roughly assessed by submitting homologous aquaporin structures to the POSA web server and examining structural differences between them. Four top FFAS hits from the PDB database can be well aligned by POSA without introducing additional flexibility which suggests that conserved common core of those structures is probably also conserved in TM0376 and the central protein core of the model is expected to be accurate.

5. *Conservation of lengths in homologous protein domains.* There is a significant variation in length conservation between protein domains. In domains without structural templates it may be the result of inaccurate domain annotation. But even in domains with known structural templates real lengths may vary significantly between homologues. Structural core of a domain is usually conserved but secondary structure elements such as α-helices and β-strands may have different lengths and be more or less extended outside the core. The most significant length variations occur in loop regions. Long loops or even entire domains may be inserted into structures of known domains changing their lengths by several hundreds of residues. Naturally, large variation of domain length makes assignment of domain boundaries more difficult, especially if sequence identity between query protein and template domain is low or alignment contains long gaps. In such cases it is recommended to check variation of domain lengths in a protein family for example by examining multiple sequence alignment available from the Pfam database (http://pfam.janelia.org/).

6. *Sequence identity as a criterion of structural similarity and alignment accuracy.* Sequence identity is the most popular and easy to understand measure of similarity between two protein sequences. Therefore, it is important to know that its value depends not only on two aligned sequences but also on the method used to align them, and on the length of the alignment. For example, a pair of homologous protein sequences aligned with FFAS will usually have longer alignment and lower sequence identity than the same pair aligned with BLAST. This is a result of the fact that a more sensitive method such as FFAS often extends the alignment to regions of lower similarity between the query and the template. Sequence identity itself is not a good criterion of sequence homology. *E*-values or *Z*-scores provide much better discrimination between similar and dissimilar pairs of protein sequences. However, if it is already known that some template sequences are similar to the query, then sequence identity is useful in picking up the most similar modeling template. If two different templates have similar alignment length with the query, then the template with higher sequence identity to the query sequence is likely to be more accurate. Sequence identity has higher correlation with structural similarity because it is related to similarity between query sequence and individual template, while sensitive sequence similarity algorithms are usually based on profile–sequence or profile–profile alignments and, therefore, they do not "see" individual sequences.

7. *Geometric feasibility of the model.* Sequence–sequence alignment is the assignment of two strings of characters which does not take into account the three-dimensional structures of proteins. In comparative modeling, the structure of the query protein is unknown, but the structure of the template structure is known and it imposes some constraints on sequence alignment. In particular, it has to be geometrically feasible to connect pairs of aligned sequence fragments (separated by gaps). Some modeling programs, such as SCWRL, which do not build unaligned regions are not sensitive to this issue, since they model only query regions aligned with the template. Other programs, such as SWISS-MODEL, will directly check geometric feasibility of the alignment and alter geometry of some aligned residues in order to connect aligned regions. If alignment is not geometrically feasible, then the model will not be built. Another group of programs, such as MODELLER, will change global geometry of the model to make connections imposed by the query–template alignment. This will usually result in significant distortions of geometry, if the alignment is geometrically infeasible.

8. *Comparison of alternative modeling templates and comparison of resulting models.* The comparison of known structures

from a protein family makes it possible to assess the accuracy of regions of a model of another protein from this family. Regions which are highly variable in known structures are also expected to be variable (and thus less accurate) in template-based models. There are two possible variants of assessing structural variability – one can use structure alignment method to align available structural templates (homologues with known structures) or build models based on those templates and align them (also using structure alignment program). The first method provides accurate overview of structural variability in a protein family of interest. The second approach allows detection of alignment inaccuracies (assuming that residues in all models are numbered consistently with the numbering of residues in the query sequence). Structurally aligned models should have corresponding residues superimposed. If structural superposition of the backbone is accurate, but residues in some models are "shifted" with respect to other models, then it usually indicates inaccuracies in sequence–template alignments.

9. *Discrepancies between full sequence of the template and ATOM records of the PDB entry.* The most common of simple technical problems encountered in comparative modeling results simply from the differences between full protein sequence of the modeling template (which is usually used in query–template alignment) and ATOM records of the corresponding PDB entry. Often, in the process of experimental structure determination coordinates of some residues are not determined. It means that some regions of the template sequence do not correspond to any atomic coordinates in the template. On the other hand, most modeling programs require 100% correspondence between the sequence used in the query–template alignment and atomic coordinates of the template. Therefore, the original query–template alignment often has to be edited before feeding it to the modeling program. Some modeling servers such as ProtMod perform automated editing of the original alignment to take such discrepancies into account. One has to note, however, that automated alignment editing may contain errors, especially when multiple fragments are missing from the template coordinates.

References

1. Chothia, C. and Lesk, A.M. (1986) The relation between the divergence of sequence and structure in proteins. *EMBO J*, **5**, 823–826.

2. Greer, J., Mollison, K.W., Carter, G.W. and Zuiderweg, E.R. (1989) Comparative modeling of proteins in the complement pathway. *Prog Clin Biol Res*, **289**, 385–397.

3. Sander, C. and Schneider, R. (1991) Database of homology-derived protein structures and the structural meaning of sequence alignment. *Proteins*, **9**, 56–68.

4. Swindells, M.B. and Thornton, J.M. (1991) Structure prediction and modelling. *Curr Opin Biotechnol*, **2**, 512–519.

5. Xiang, Z. (2006) Advances in homology protein structure modeling. *Curr Protein Pept Sci*, 7, 217–227.

6. Ginalski, K. (2006) Comparative modeling for protein structure prediction. *Curr Opin Struct Biol*, 16, 172–177.

7. Murzin, A.G., Brenner, S.E., Hubbard, T. and Chothia, C. (1995) SCOP: a structural classification of proteins database for the investigation of sequences and structures. *J Mol Biol*, 247, 536–540.

8. Finn, R.D., Tate, J., Mistry, J., Coggill, P.C., Sammut, S.J., Hotz, H.R., Ceric, G., Forslund, K., Eddy, S.R., Sonnhammer, E.L. et al. (2008) The Pfam protein families database. *Nucleic Acids Res*, 36, D281–D288.

9. Bru, C., Courcelle, E., Carrere, S., Beausse, Y., Dalmar, S. and Kahn, D. (2005) The ProDom database of protein domain families: more emphasis on 3D. *Nucleic Acids Res*, 33, D212–D215.

10. Mulder, N.J., Apweiler, R., Attwood, T.K., Bairoch, A., Bateman, A., Binns, D., Bork, P., Buillard, V., Cerutti, L., Copley, R. et al. (2007) New developments in the InterPro database. *Nucleic Acids Res*, 35, D224–D228.

11. Gough, J., Karplus, K., Hughey, R. and Chothia, C. (2001) Assignment of homology to genome sequences using a library of hidden Markov models that represent all proteins of known structure. *J Mol Biol*, 313, 903–919.

12. Cheng, J., Sweredoski, M. and Baldi, P. (2006) DOMpro: protein domain prediction using profiles, secondary structure, relative solvent accessibility, and recursive neural networks. *Data Mining and Knowledge Discovery*, 13, 1–10.

13. Cheng, J. (2007) DOMAC: an accurate, hybrid protein domain prediction server. *Nucleic Acids Res*, 35, W354–W356.

14. Linding, R., Russell, R.B., Neduva, V. and Gibson, T.J. (2003) GlobPlot: Exploring protein sequences for globularity and disorder. *Nucleic Acids Res*, 31, 3701–3708.

15. Marsden, R.L., McGuffin, L.J. and Jones, D.T. (2002) Rapid protein domain assignment from amino acid sequence using predicted secondary structure. *Protein Sci*, 11, 2814–2824.

16. Liu, J. and Rost, B. (2004) CHOP: parsing proteins into structural domains. *Nucleic Acids Res*, 32, W569–W571.

17. Dunbrack, R.L., Jr. (2006) Sequence comparison and protein structure prediction. *Curr Opin Struct Biol*, 16, 374–384.

18. Holm, L., Ouzounis, C., Sander, C., Tuparev, G. and Vriend, G. (1992) A database of protein structure families with common folding motifs. *Protein Sci*, 1, 1691–1698.

19. Altschul, S.F., Madden, T.L., Schaffer, A.A., Zhang, J., Zhang, Z., Miller, W. and Lipman, D.J. (1997) Gapped BLAST and PSI-BLAST: a new generation of protein database search programs. *Nucleic Acids Res*, 25, 3389–3402.

20. Eddy, S.R. (1998) Profile hidden Markov models. *Bioinformatics*, 14, 755–763.

21. Rychlewski, L., Jaroszewski, L., Weizhong, L. and Godzik, A. (2000) Comparison of sequence profiles. Structural predictions with no structure information. *Protein Sci*, 8, 232–241.

22. Soding, J. (2005) Protein homology detection by HMM-HMM comparison. *Bioinformatics*, 21, 951–960.

23. Chandonia, J.M., Hon, G., Walker, N.S., Lo Conte, L., Koehl, P., Levitt, M. and Brenner, S.E. (2004) The ASTRAL Compendium in 2004. *Nucleic Acids Res*, 32, D189–D192.

24. Altschul, S.F., Gish, W., Miller, W., Myers, E.W. and Lipman, D.J. (1990) Basic local alignment search tool. *J Mol Biol*, 215, 403–410.

25. Kelley, L.A., MacCallum, R.M. and Sternberg, M.J. (2000) Enhanced genome annotation using structural profiles in the program 3D-PSSM. *J Mol Biol*, 299, 499–520.

26. Shi, J., Blundell, T.L. and Mizuguchi, K. (2001) FUGUE: sequence-structure homology recognition using environment-specific substitution tables and structure-dependent gap penalties. *J Mol Biol*, 310, 243–257.

27. Fischer, D. (2000) Hybrid fold recognition: combining sequence derived properties with evolutionary information. *Pac Symp Biocomput*, 119–130.

28. Xu, Y. and Xu, D. (2000) Protein threading using PROSPECT: design and evaluation. *Proteins*, 40, 343–354.

29. Karplus, K., Barrett, C. and Hughey, R. (1998) Hidden Markov models for detecting remote protein homologies. *Bioinformatics*, 14, 846–856.

30. Jaroszewski, L., Rychlewski, L. and Godzi, A. (2000) Improving the quality of twilight-zone alignments. *Protein Sci*, 9, 1487–1496.

31. Durbin, R., Eddy, S., Krogh, A. and Mitchison, G. (1998) *Biological Sequence Analysis: Probabilistic Models of Proteins and Nucleic Acids*. Cambridge University Press: Cambridge.

32. Krogh, A., Larsson, B., von Heijne, G. and Sonnhammer, E.L. (2001) Predicting transmembrane protein topology with a hidden Markov model: application to complete genomes. *J Mol Biol*, 305, 567–580.

33. Lupas, A., Van Dyke, M. and Stock, J. (1991) Predicting coiled coils from protein sequences. *Science*, **252**, 1162–1164.

34. Ward, J.J., Sodhi, J.S., McGuffin, L.J., Buxton, B.F. and Jones, D.T. (2004) Prediction and functional analysis of native disorder in proteins from the three kingdoms of life. *J Mol Biol*, **337**, 635–645.

35. Jones, D.T. (1999) Protein secondary structure prediction based on position-specific scoring matrices. *J Mol Biol*, **292**, 195–202.

36. Wootton, J. and Federhen, S. (1993) Statistics of local complexity in amino acid sequences and sequence databases. *Comput Chem*, **17**, 149–163.

37. Ginalski, K. and Rychlewski, L. (2003) Detection of reliable and unexpected protein fold predictions using 3D-Jury. *Nucleic Acids Res*, **31**, 3291–3292.

38. Sanchez, R. and Sali, A. (1997) Advances in comparative protein-structure modelling. *Curr Opin Struct Biol*, 7, 206–214.

39. Wallner, B. and Elofsson, A. (2005) All are not equal: a benchmark of different homology modeling programs. *Protein Sci*, **14**, 1315–1327.

40. Michalsky, E., Goede, A. and Preissner, R. (2003) Loops In Proteins (LIP) – a comprehensive loop database for homology modelling. *Protein Eng*, **16**, 979–985.

41. Xiang, Z., Soto, C.S. and Honig, B. (2002) Evaluating conformational free energies: the colony energy and its application to the problem of loop prediction. *Proc Natl Acad Sci USA*, **99**, 7432–7437.

42. Sali, A. (1994) Modeller. A program for protein structure modelling by satisfaction of spatial restraints. http://quitar.rockefeller.edu/modeller/modeller.html.

43. Canutescu, A.A., Shelenkov, A.A. and Dunbrack, R.L., Jr. (2003) A graph-theory algorithm for rapid protein side-chain prediction. *Protein Sci*, **12**, 2001–2014.

44. Vriend, G. (1990) WHAT IF: a molecular modeling and drug design program. *J Mol Graph*, **8**, 52–56, 29.

45. Schwede, T., Kopp, J., Guex, N. and Peitsch, M.C. (2003) SWISS-MODEL: an automated protein homology-modeling server. *Nucleic Acids Res*, **31**, 3381–3385.

46. Reeves, G.A., Dallman, T.J., Redfern, O.C., Akpor, A. and Orengo, C.A. (2006) Structural diversity of domain superfamilies in the CATH database. *J Mol Biol*, **360**, 725–741.

47. Ye, Y. and Godzik, A. (2005) Multiple flexible structure alignment using partial order graphs. *Bioinformatics*, **21**, 2362–2369.

48. Bowie, J.U., Luthy, R. and Eisenberg, D. (1991) A method to identify protein sequences that fold into a known three-dimensional structure. *Science*, **253**, 164–170.

49. Sippl, M.J. (1993) Recognition of errors in three-dimensional structures of proteins. *Proteins*, **17**, 355–362.

50. Morris, A.L., MacArthur, M.W., Hutchinson, E.G. and Thornton, J.M. (1992) Stereochemical quality of protein structure coordinates. *Proteins*, **12**, 345–364.

51. Hooft, R.W., Vriend, G., Sander, C. and Abola, E.E. (1996) Errors in protein structures. *Nature*, **381**, 272.

52. Melo, F., Devos, D., Depiereux, E. and Feytmans, E. (1997) ANOLEA: a www server to assess protein structures. *Proc Int Conf Intell Syst Mol Biol*, **5**, 187–190.

53. Word, J.M., Lovell, S.C., LaBean, T.H., Taylor, H.C., Zalis, M.E., Presley, B.K., Richardson, J.S. and Richardson, D.C. (1999) Visualizing and quantifying molecular goodness-of-fit: small-probe contact dots with explicit hydrogen atoms. *J Mol Biol*, **285**, 1711–1733.

54. Jaroszewski, L., Rychlewski, L., Li, Z., Li, W. and Godzik, A. (2005) FFAS03: a server for profile – profile sequence alignments. *Nucleic Acids Res*, **33**, W284–W288.

55. Wallner, B. and Elofsson, A. (2005) Pcons5: combining consensus, structural evaluation and fold recognition scores. *Bioinformatics*, **21**, 4248–4254.

56. Fischer, D. (2003) 3D-SHOTGUN: a novel, cooperative, fold-recognition meta-predictor. *Proteins*, **51**, 434–441.

57. Fischer, D., Rychlewski, L., Dunbrack, R.L., Jr., Ortiz, A.R. and Elofsson, A. (2003) CAFASP3: the third critical assessment of fully automated structure prediction methods. *Proteins*, **53**(Suppl 6), 503–516.

58. Rychlewski, L. and Fischer, D. (2005) LiveBench-8: the large-scale, continuous assessment of automated protein structure prediction. *Protein Sci*, **14**, 240–245.

59. Fischer, D. (2006) Servers for protein structure prediction. *Curr Opin Struct Biol*, **16**, 178–182.

60. Rost, B., Yachdav, G. and Liu, J. (2004) The PredictProtein server. *Nucleic Acids Res*, **32**, W321–W326.

61. McGuffin, L.J., Bryson, K. and Jones, D.T. (2000) The PSIPRED protein structure prediction server. *Bioinformatics*, **16**, 404–405.

62. Pieper, U., Eswar, N., Davis, F.P., Braberg, H., Madhusudhan, M.S., Rossi, A., Marti-Renom, M., Karchin, R., Webb, B.M., Eramian, D. et al. (2006) MODBASE: a database of annotated

comparative protein structure models and associated resources. *Nucleic Acids Res*, **34**, D291–D295.

63. Eswar, N., John, B., Mirkovic, N., Fiser, A., Ilyin, V.A., Pieper, U., Stuart, A.C., Marti-Renom, M.A., Madhusudhan, M.S., Yerkovich, B. et al. (2003) Tools for comparative protein structure modeling and analysis. *Nucleic Acids Res*, **31**, 3375–3380.

64. Bates, P.A., Kelley, L.A., MacCallum, R.M. and Sternberg, M.J. (2001) Enhancement of protein modeling by human intervention in applying the automatic programs 3D-JIGSAW and 3D-PSSM. *Proteins*, Suppl 5, 39–46.

65. Slabinski, L., Jaroszewski, L., Rychlewski, L., Wilson, I.A., Lesley, S.A. and Godzik, A. (2007) XtalPred: a web server for prediction of protein crystallizability. *Bioinformatics*, **23**, 3403–3405.

66. Fernandez-Fuentes, N., Rai, B.K., Madrid-Aliste, C.J., Fajardo, J.E. and Fiser, A. (2007) Comparative protein structure modeling by combining multiple templates and optimizing sequence-to-structure alignments. *Bioinformatics*, **23**, 2558–2565.

Chapter 8

Applications of Bioinformatics to Protein Structures: How Protein Structure and Bioinformatics Overlap

Gye Won Han, Chris Rife, and Michael R. Sawaya

Summary

In this chapter, we will focus on the role of bioinformatics to analyze a protein after its protein structure has been determined. First, we present how to validate protein structures for quality assurance. Then, we discuss how to analyze protein–protein interfaces and how to predict the biomolecule which is the biological oligomeric state of the protein. Finally, we discuss how to search for homologs based on the 3-D structure which is an essential step for understanding protein function.

Key words: Protein structure validation, Biomolecule, Protein interaction, 3-D structure homologs, Bioinformatics

1. Introduction

The total number of macromolecular structures in the Protein Data Bank (PDB, http://www.rcsb.org) *(1)* continues to grow. As of February 5, 2008, 48,700 structures have been deposited in the PDB. Most are determined by X-ray crystallography (85%) and NMR (14%) with electron microscopy and other methods (1%) accounting for the remaining structures. The number of PDB depositions per year is also growing. Last year, more than 7,200 structures were deposited in the PDB, which is more than double that of 2002.

The number of protein structures deposited by the Structural Genomics (SG) centers has also been steadily increasing each year. SG is a worldwide initiative aimed at quickly determining a large number of protein structures using X-ray crystallographic

Vadim Astakhov (ed.), *Biomedical Informatics,* Methods in Molecular Biology, vol. 569
DOI 10.1007/978-1-59745-524-4_8, © Humana Press, a part of Springer Science+Business Media, LLC 2009

and NMR technologies. Protein Structure Initiative (PSI) SG centers were established in 2000 by the National Institute of General Medical Sciences (NIGMS) *(2)*. During the PSI-1 period, from 2000 to 2005, approximately 1,300 structures were determined by nine PSI SG centers. In PSI-2, from July 2005 through February 2008, a total of 1,414 structures were deposited by PSI SG centers. The four large-scale production centers accounted for 1,310 of those structures.

In addition to their focus on protein structure determination, PSI SG centers also contribute significantly to technology development, including high-throughput techniques, remote data collection, and advances in data processing and refinement. The Quality Control Check program discussed later in this chapter was developed at the Joint Center for Structural Genomics (JCSG) (http://www.jcsg.org).

In this chapter, we will discuss in detail what steps follow the determination of a protein structure. This includes validating the protein structure, calculation of the most probable grouping of molecules that forms the biomolecular assembly, and search for structural homologs by the use of recently developed bioinformatics programs. The biomolecule calculation can be useful in determining the oligomeric state of the molecule, while the structural similarity search can help in determining the function of a protein of unknown function.

2. Validation of the Protein Structure

While the challenges of protein crystallization and phase determination are well described, less attention has been given to the point at which a structure can be considered finished and how to most efficiently get to that point. In order to ensure that completed structures are of consistently high quality, it is helpful to apply a set of standard target values and to regularly track the progress of refinement in relation to those targets. During this quality control (QC) step, various aspects of the model and associated files should be inspected manually and with currently available programs. As a means of making the QC process more efficient, these checks have been incorporated into a web-based interface. These automated checks include analyzing the refinement statistics, checking the formatting of the final PDB file *(1)*, and evaluating the model for any problem areas. Upon completion of a run, the QC script will return a report to the user, enabling them to quickly judge which areas of the structure, if any, need further attention.

2.1. Running the QC Script

Running the QC script is accomplished via an interface available at http://smb.slac.stanford.edu/~crife/qc_check/qc_check.html[1]. The user is prompted to upload up to four different files. The protein coordinate file *(Pdb_file)*, structure factor file *(mtz_file)*, and REFMAC5 *(3)* refinement log file *(refmac_log)* are the most recent files from the current round of refinement, and *sequence* is the sequence file in FASTA format. Currently, the script works exclusively on structures refined using the program REFMAC5 *(3)* from the CCP4 suite *(4)*. There are a number of checks available, and the user is requested to select which jobs they would like to run. It is only necessary to upload the files that correspond to the selected job(s) requested. For example, selecting Molprobity *(5)* requires only the PDB file, while the sequence check requires the PDB and sequence files. Most browsers will list the required input files if the mouse is held over the check-boxes.

2.2. Description of Checks

2.2.1. Refinement Statistics and PDB Check

This check requires the PDB, mtz, and REFMAC5 log files. During the check of the refinement statistics, some basic information from the log file is summarized, and the values for R_{work}, R_{free}, calculated expectation values *(6, 7)*, map correlation, RMS bond length, and chiral-center are listed, and the values are highlighted if they exceed predefined limits. Some refinement parameters are also displayed and checked, such as the value of the phase restraints (if used), B-factor restraints, presence of riding hydrogens, use of noncrystallographic symmetry (NCS), the number of TLS groups used, and if there are any MAKE_U_POSITIVE errors during the TLS refinement. These checks examine the structure refinement statistics to look for potential problems during the refinement.

The PDB check uses Moleman2 *(8)* to look for any atoms that are not at full occupancy, with methionine and seleno-methionine being listed separately. The highest and lowest B values are listed, and any atoms with values of 2.00 are flagged. Finally, the protein and nonprotein chain IDs are presented, and the PDB file is checked for the presence of REMARK 300, 350, and 999, although the correctness of those remarks is not evaluated.

2.2.2. Sequence Check

The sequence check requires the PDB file and the sequence in FASTA format. Each protein chain in the PDB is checked against the sequence using CLUSTALW *(9)*, and the full alignment is presented. Mismatches are highlighted in the alignment and listed separately for clarity, and the residues from any gaps are also listed.

---

[1] The Quality Control Check program was developed at the Joint Center for Structural Genomics (JCSG). If you use this tool in preparing a structure please acknowledge the JCSG.

2.2.3. NCS Check

This requires only the PDB file. If there are multiple copies of the protein in the asymmetric unit (ASU), the NCS check can be used to evaluate any differences between the residues of each chain. All the chains are aligned using superpose *(10)*, and the central chain is determined based on the RMS differences between the chains. Simple "flips" are listed first, such as corresponding threonines having their OG1 and CG1 atoms reversed, followed by instances where the residues are in moderately and substantially different positions. Additionally, a suggested NCS restraints definition is generated for use in REFMAC. A PDB file with the aligned protein chains, along with a Coot *(11)* script are available for download. This check was developed by Abhinav Kumar (personal communication).

2.2.4. Molprobity Check

Molprobity is used to evaluate the geometry of the model. The REMARK 42 table generated by Molprobity is displayed, and any residues flagged as Ramachandran outliers, rotamer outliers, or possible hydrogen-bond flips are listed. The Coot script generated by Molprobity can be downloaded to facilitate easier editing of the structure. This requires only the PDB file.

2.2.5. ADIT Check

The PDB's AutoDep Input Tool (ADIT) provides a useful means for evaluating many parts of the model, including intermolecular solvent distances, close contacts, covalent bond lengths and angles, as well as torsion angles and chirality. Any problem areas are listed in the output. These checks utilize NUCHECK *(12)*, PROCHECK *(13)*, and SFCHECK *(14)*, and only require the PDB file.

2.2.6. Resolve Check

The final check available is resolve_evaluate *(15)*. Resolve uses the PDB and mtz files to calculate an overall map correlation and determine the number of residues in acceptable, weak, and out of density. Residues in questionable density are listed for further review.

2.3. Presenting the Results

Once the run is completed, all results are output to the screen. Hyperlinks are provide in order to quickly navigate to a particular check. Under the "Input Comments and Save Results" section (located at the bottom of the page), there is a box in which the user can optionally add comments in a free text format. Once any comments have been added, the entire output can then be saved as an html file using the "Save comments and QC output" box. The file will be saved, even if no comments are entered.

2.4. Using the QC Check

Although the checks provided above collect many useful programs and organize their output for easy dissemination, ultimately the user must still determine when the structure is completed. The suggested values and ranges used by the program are general suggestions and may not be ideal for every refinement scenario. However, if used throughout the refinement, there is a significant time savings obtained by running the QC script versus running each job individually and manually sorting through the results. Over the course of many cycles of refinement and multiple structures, this will enable the crystallographer to become more efficient.

3. Biomolecules and the Protein–Protein Interface

The protein coordinate files in the PDB contain only the asymmetric unit (ASU) of a crystal. The ASU is the smallest unit that can be used to build a structure in the unit cell by crystal symmetry operations. The biomolecule (biological molecule), which is the protein's native oligomeric state, is often different than the number of molecules found in the ASU. The biomolecule can be one ASU, several ASUs, or parts of adjoining ASUs.

Explanation of the crystallographic ASU, including the number of biomolecules described by the PDB coordinates, is listed in the REMARK 300 section of the PDB header. The information on generating the biological molecule by both crystallographic and noncrystallographic operations is listed in the REMARK 350.

The PDB provides separate coordinate and image files containing the biological molecule. These can be accessed from the Structure Summary "Images and Visualization" and "Download Files" pages at the PDB web page.

Sometimes, it is difficult to discern whether a grouping of molecules represents a true biological assembly or is simply a crystal packing artifact derived from crystallographic and noncrystallographic symmetry. Therefore, it is recommended to confirm the biologically relevant oligomeric state using size-exclusion liquid chromatography and static light scattering experiments.

3.1. Protein Interfaces, Surface and Assemblies Service (PISA) (16)

PISA server is the most frequently used web server, which provides various analyses of protein structures based on physical–chemical models of macromolecular interactions and chemical thermodynamics. It calculates protein surfaces, significance of protein interfaces (biological role), hydrogen bonds, salt bridges, disulfide bonds, covalent bonds, and prediction of probable quaternary structures (assemblies). Similar analyses are performed by the Protein Quaternary Structure (PQS) server at MSD-EBI (17) and PITA software (18) (**Table 1**). In addition, PISA can also detect the ligands bound in the PDB file and let the user decide whether to include them in the analysis. This option is very useful to eliminate interactions (e.g., solvent-mediated) which are not relevant to the true protein–protein interaction. Servers that calculate the protein–protein interaction and interface analysis of structures, either designed by PDB accession code or uploaded by the user, are listed in **Table 1**.

Table 1
Comparison of protein–protein interaction servers

Program/reference/URL	Summary	Output presentation
FastContact (19) http://structure.pitt.edu/serv-ers/fastcontact/	Useful only for protein–protein complex. Needs receptor and ligand PDB files	E-mail only
InterProSurf (20) http://curie.utmb.edu/pdb-complex.html	Accepts PDB accession code – the results are reliable for only multichain PDB file	Web page only. E-mail acknowl-edgment
PIC (21) http://crick.mbu.iisc.ernet.in/~PIC/	Accepts PDB accession code as well as PDB coordinates which contains mul-tichain protein	Web page only
PISA (16) http://www.ebi.ac.uk/msd-srv/prot_int/pistart.html	Accepts PDB accession code as well as PDB coordinates. It predicts biomol-ecule assemblies and also provides the detailed protein–protein interaction between biomolecule	Web page only. The biomolecule coor-dinates available for download
PITA (18) http://www.ebi.ac.uk/thornton-srv/databases/pita/	Accepts PDB accession code as well as PDB coordinates to predict biological unit for a protein and a method of scor-ing each trial protein–protein interface	Web page only. The biomolecule coor-dinates available for download
PPI (22) http://www.biochem.ucl.ac.uk/bsm/PP/server/	Reliable results can be obtained only when multichain PDB file is uploaded	E-mail is sent with link to a web page to display results
PQS (17) http://pqs.ebi.ac.uk/	Calculates protein quaternary structures based on the crystal symmetry opera-tion. Accepts PDB accession code only	Web page only. The biomolecule coor-dinates available for download
PROFACE (23) http://202.141.148.29/resources/bioinfo/interface/	Input PDB file should contain multichains	Web page only
SPPIDER (24) http://sppider.cchmc.org/	The program identifies the interface within known protein–protein complex either in ASU or biological unit	E-mail and web page output

3.2. FastContact (19)

FastContact 2.0 is a free energy scoring tool for protein–protein structures. It requires "receptor PDB" as well as "ligand PDB" files. This is useful for calculating protein–protein interactions of the known complex.

3.3. InterProSurf (20)

InterProSurf calculates the residues present at the protein interface, polar area, apolar area, number of surface atoms, number of buried atoms, and total area of the molecule in the protein complex as well as in its monomer form. It needs only the PDB accession code, and the server automatically detects the number of chains present in the PDB file and outputs the results for all chains present in the protein complex.

3.4. Protein Interactions Calculator (PIC) (21)

PIC is a server that identifies various kinds of interactions, such as disulfide bonds, hydrophobic interactions, ionic interactions, hydrogen bonds, aromatic–aromatic interactions, aromatic–sulfur interactions and cation–pi interactions within a protein or between proteins in a complex. It also determines the accessible surface area, as well as the distance of a residue from the surface of the protein. The input should be in the PDB format (both PDB accession code as well as PDB coordinates are accepted). Interactions are calculated based on empirical or semiempirical sets of rules.

3.5. Protein–Protein Interaction Server (PPI) (v1.5) (22)

The PPI server is a tool to analyze the protein–protein interface of any protein complex. It requires the PDB coordinates of a protein structure. The analyzed results are tabulated with the nature of the protein–protein interface including accessible surface area, planarity, length and breadth, secondary structure, hydrogen bonds, salt bridges, gap volume, gap volume index, bridging water molecules, and interface residues.

3.6. ProFace (23)

ProFace is a suite of programs that use a file containing atomic coordinates of a multichain molecule as input and analyzes the interface between any two or more subunits. The interface residues are shown segregated into spatial patches (if such a clustering is possible based on an input threshold distance) and/or core and rim regions. A number of physicochemical parameters defining the interface are tabulated. Among the different output files, one contains the list of interacting residues across the interface. Results can be used to infer if a particular interface belongs to a homodimeric molecule. ProFace requires the Java Virtual Machine (JVM) to be installed to view the Java applets from the server.

3.7. Solvent Accessibility Based Protein–Protein Interface iDEntification and Recognition (SPPIDER) (24)

SPPIDER can predict interaction sites using a structure of a single protein chain or known complexes. This server can also calculate protein–protein interactions based on the biological unit. Two query options are available in this server – ASU or PQS biological unit – for calculation of the protein–protein interaction.

To demonstrate how the protein–protein interaction and interface analysis programs work, we present three test cases.

1. Alanine–glyoxylate aminotransferase (ALR1004) from *Anabaena* sp. (PDB accession code: 1vjo). ALR1004 is a monomer in the ASU and the biological molecule is a dimer *(25)*. PISA, PQS, and PITA results agree on a dimer as the molecular assembly. PISA also suggests a detailed interface with the surface and buried area as well as *G* (solvation free energy, kcal/M). SPPIDER also predicts the interaction site (interacting residues). The other protein interaction servers listed here are expected to provide the detailed information of the protein–protein interaction in the dimer. However, either the input monomer was not acceptable or the program incorrectly identified some residues as involved in an interface when they were quite far away.

2. Apo mRNA decapping enzyme (DcpS) from *Mouse* (PDB accession code: 1vlr). DcpS is a dimer in the ASU and also a dimer as a biological molecule *(26)*. PISA, PQS, and PITA agreed that the biological assembly is a dimer. Other protein interaction servers listed in **Table 1** provided the detailed information of the protein–protein interaction of the dimer.

3. TM0067 (PDB accession code: 2afb). TM0067 contains a dimer in the ASU and the biomolecule is a hexamer *(27)*. PISA, PQS, and PITA predict a hexamer as the biomolecule and provide the detailed information of all interactions in the hexamer. SPPIDER can find the interacting residues based on the PQS biological unit when the PQS query was selected in the query. However, all other programs/servers listed in **Table 1** provided the output of the dimer interaction between two chains without taking crystallographic symmetry into account.

PISA (and PQS and PITA) and SPPIDER are particularly useful if crystallographic symmetry contained in the PDB file is important to the oligomeric state of the protein/complex. Programs other than PISA and SPPIDER can be used for the analysis of the protein–protein interactions but require the proper input PDB file to account for the crystal packing. Therefore, use of the multimer PDB coordinates downloaded from the PISA server for calculating the detailed protein–protein interaction is recommended, rather than submitting the PDB accession code.

However, if crystal symmetry is not an issue, the analysis of the protein–protein interactions can be obtained using any one of the **Table 1** programs depending on the type of analysis required. All listed servers and programs perform the analysis in a reasonable amount of time (usually a few minutes).

4. Homolog Search Based on 3-D Protein Structures

A structural biologist can uncover a wealth of information about a protein molecule by ascertaining whether it shares structural similarity with a molecule already archived in the PDB. This search for structural homologs is informative for several reasons: (1) to determine whether a protein fold is novel; (2) to draw evolutionary relationships between proteins and organisms; (3) to infer a function for the query molecule from the established function of the archived molecule. This last point has become increasingly important, given the long-term objective of structural genomics groups to determine representative structures of all protein families encoded by a host's genome *(28)*. Now, more than ever, a structural biologist is likely to have obtained a protein structure with little or no prior evidence of its function.

Today, dozens of web servers are available to assist the structural biologist in finding a structural homolog (**Table 1**). These web servers differ in speed, sensitivity, specificity, search algorithms, ranking/scoring of results, graphical display capabilities, aligned coordinate output capability, level of maintenance, and frequency of database updates. The best choice of web server is dependent on the goals and capabilities of the user. To help choose the most suitable web servers, some of the key differences in search algorithms and performance are discussed and compared with an example.

Structural alignment servers differ most significantly in three categories: sensitivity, specificity, and speed. In theory, brute force would provide for the most sensitive and specific search results, but would require weeks or more of computation given the complexity of protein structure and the presence of over 50,000 structure entries in the PDB. To shorten the search time, the authors of these structural alignment search tools have devised simplifying representations of protein structure. These representations can be aligned faster than full atom representations. The choice of which characteristics of protein structure become encoded in the representation is guided by the author's primary purpose: sensitivity, specificity, or speed. For example, the DALI algorithm represents protein structure as a matrix of interatomic distances *(29)* that retain many of the details of the protein structures. This algorithm produces one of the most sensitive and specific structural similarity searches, but also one of the lengthiest, typically completing in about 4 h for a domain of approximately 100 residues. At the other end of the spectrum, the TOPSCAN algorithm represents protein more simply as a collection of secondary structure elements, each reduced to a vector encoding direction, proximity, accessibility, and length *(30)*. A complete search with TOPSCAN requires only 30 s and reveals topological homologs that DALI would not find, but at a higher risk of encountering false positives (i.e., good sensitivity, poor specificity, good speed).

Given the range of sensitivity and specificity encountered, it is clearly important to be able to discern the significance of a structural alignment. The most intuitive method is to evaluate the length of the aligned regions and the RMSD between Cαs in the alignment. Longer lengths and lower RMSDs suggest a significant alignment. For example, an alignment covering 100 amino acids or more might be considered significant if the RMSD were under 3.5 Å. Shorter alignment regions would require smaller RMSDs to be considered significant. A more convenient measure of the significance of a structural alignment is provided by the probability score, that is, the probability that a match of a given quality would be obtained by choosing any pair of structures at random. An example is the P-score given by VAST (31), where a score of 0.0001 would be considered significant. Many web servers offer both types of scoring systems (**Table 2**). Some offer the flexibility of re-sorting the matches by different scoring systems. Usually, the rank orders imposed by the different scoring schemes are well correlated. Exceptions occur when the search model contains a flexible hinge. If the aligned pair of structures differs in hinge angle, a large RMSD will result (e.g., ~7.0 Å), but the match can still be significant if indicated by the P-score.

There are additional matters on convenience worth considering such as choice of target database, and capability of viewing the aligned structures on-the-fly. Table 2 reports these qualities for ten web servers tested. Some web servers provide aligned coordinates and the graphics plugin to view them. Others simply provide the name of the matched PDB file and it is up to the user to perform the superposition. The target structural database used can vary. The usual choice of target database is a subset of the PDB in which structures with sequence identities of over 45% have been removed. The use of a subset, rather than the entire PDB can speed up the structural search and subsequent analysis by the user. It is also important to consider that these databases vary in the frequency of updates; a more frequently updated web server might reveal a structural match not even considered by other web servers.

To test the quality of the different web servers, the structure of PduU (32), a structural component of the ethanolamine utilization pathway, was submitted to all ten web sites. The PduU structure is a circularly permuted variant of a newly described protein fold called the biological micro-compartment (BMC) fold (33). PduU shares significant similarity with several structures in the PDB (e.g., PDB accession code 2g13, 2a1b, 2ewh, 2a18, and 2a10; RMSD for 74 Cα pairs is under 1.4 Å), but the circular permutation in PduU means that when compared with other members of the BMC family structurally equivalent segments differ in the order in which they are connected. Some of the web servers were able to recognize the structural similarity in spite of topological differences (DALI, MATRAS, and FATCAT). But many of the web servers proved to be sensitive to the difference in connectivity and were, therefore, incapable of recognizing the strong relationship to the BMC fold (e.g., VAST, SSM, TOPSCAN, PRIDE, and ProteinDBS) (**Table 2**). This blindness to circular permutants is another factor to consider when choosing a structural alignment web server.

In summary, the appropriate choice of structure alignment server depends on the goal of the user. If the user wishes to find only very closely related structures (i.e., RMSD of 1.5 Å for 100 superimposed Cαs), then almost any of these servers will suffice. The user could simply pick the speediest web server with the most attractive display/output. But if the goal is to infer a function for a protein from the function of structural homologs, it would be best to use a few different web servers to ensure that no significant match was missed. In a favorable case, a majority of the potential homologs will share a common function. The DALI server is the most frequently used server and provides high degrees of sensitivity and selectivity, though the search time is the longest of those tested (4 h). The MATRAS *(34)* and FATCAT *(35)* servers provide sensitivity and selectivity comparable to DALI but at faster speeds (20 min or 2 h, respectively). FATCAT provides the additional attractive feature of an interactive graphics display to view the structural alignment on request.

5. Notes and Conclusion

When presented with the structure of a protein of unknown function, the reader should now be knowledgeable about a variety of web-based tools to infer the function of the protein. These tools enable the reader to (1) validate protein structures, (2) determine the molecule's biological oligomerization state, and (3) uncover homologs based on 3-D structure. In so doing, the reader will have uncovered valuable clues hidden in the protein's structure. This information can then be used to help plan subsequent functional biochemical experiments, such as mutation analyses.

Table 2
Comparison of structural alignment servers

Program name/reference/ algorithm/URL	Search time/found BCM domain?	Target options	Output presentation/output scoring options	Top 10 matches (PDB accession codes)
Dali Server (29) Distance matrix alignment http://www.ebi.ac.uk/dali/	4 h/Y	PDB representatives	E-mail only. Matrices given for superposition./Z-score, RMSD, aligned length, % sequence identity	2a10, 2hmf, 2dtj, 1u8s, 1hi9, 2d6f, 1xxa, 1mia, 2phc, 2cz4
CE (36) Combinatorial extension http://cl.sdsc.edu/ce/all-to-all/1-to-all.html	Not operational at time of testing	All PDB (complete) or representatives (fast)	E-mail only. Sequence alignment available. Aligned coordinates not available./Z-score, RMSD, aligned length, % sequence identity	Not operational at time of testing
VAST (31) Vector alignment http://www.ncbi.nlm.nih. gov/Structure/VAST/ vastsearch.html	4 min/N	All PDB (complete) or medium redundancy (fast)	Web page only. Options for viewing structural (Cn3D) and sequence alignments graphically./Vast score, P score, RMSD, aligned length, % sequence identity, and aligned length	By VAST score 1yl1, 2re1, 1fx2, 2eua, 1fx4, 1cul, 1f9n, 1f6f
SSM (10) Secondary structure matching http://www.ebi.ac.uk/msd-srv/ssm/cgi-bin/ssmserver	30 s/N	Choice of PDB or SCOP or subsets thereof. Choice of connectivity constraint	Web page only. Alignments downloadable. Options for viewing structural (Jmol or Rasmol) and sequence alignments graphically./Q score, P score, Z score, RMSD, aligned length, % sequence identity, aligned SSE, number of gaps	1xxa, 1b4b, 2p5m, 2ril, 2hiq, 2hgm
TOPSCAN (30) Alignment of secondary topology strings http://www.bioinf.org.uk/ topscan/	30 s/N		Web page only. Links to PDBSUM. No alignments given./Topscan score (%)	2vhi, 2kfn, 1qap, 1mml, 1tig, 1mla, 1ris, 1mxa, 1cdw, 1apy

Name / description / URL	Time/Y-N	Database	Output	PDB codes
MATRAS (34) SSE alignment, similarity score based on log odds formula http://biunit.naist.jp/matras/	19 min/Y	PDB or SCOP representatives	E-mail only. Sequence alignments given. No coordinate output./Z-score, % sequence identity, SSE score	2ewh, 1qd1, 2pfd, 1l0w, 1xmb, y10, 1ybt, 1bia, 1u8s, 1wc3
PRIDE (37) Compares Cα–Cα distance distributions http://hydra.icgeb.trieste.it/pride/	10 s/N	PDB-select or CATH-select	Web page only. No sequence or coordinates output./PRIDE2 score	1kcq, 1pil, 1o7b, 1s4z, 2bop, 2xn7, 1wz3, 1o5l, 1mhm, 2agh
FATCAT (35) Allows alignments between proteins with different conformational states http://fatcat.burnham.org/	2 h/Y	PDB or CATH or subsets thereof	E-mail is sent with link to a web page to display results. Structural alignment viewed with (JMOL or CHIME) and sequence alignments available. Aligned coordinates available for download./FATCAT score, P-value, optimized RMSD, aligned length	2aib, 2a10, 2wwh, 2eky, 2ibo, 1yqh, 1fe0, 1vk8, 1lxn, 1usm
ProteinDBS (38) Matches the patterns of 2-D distance matrices with computer vision algorithms http://proteindbs.rnet.missouri.edu/PDBS_V2.php	20 s/N	All PDB	Web page only. Structural alignment viewed with Kinemage. Sequence alignments given./ProteinDBS ranking. RMS deviation. Alignment length given	1i94, 1emw, 1hnz, 2awt, 2ca4, 1wf2, 2bpb
Cathedral Server (39) http://www.cathdb.info/cgi-bin/cath/CathedralServer.pl	7 min/N	CATH	Web page only. No coordinates or alignment available./Cathedral score, RMSD, E-value	1mnn, 1jz7, 1gtf, 1gme, 1n01, 1twf, 1ru0, 1dar, 1iq0, 1dqi

Acknowledgment

G.W.H. and C.R. are affiliated with the JCSG, which is supported by the National Institute of General Medical Sciences, Protein Structure Initiative; Grant Numbers P50 GM62411, U54 GM074898. M.R.S. is supported by the Howard Hughes Medical Institute. We thank Ian Wilson, Anand Kolatkar, and Robyn Stanfield for their comments on this chapter. We also thank Lisa Van Veen for technical assistance. This is TSRI manuscript number 19238. The content is solely the responsibility of the authors and does not necessarily represent the official views of the National Institute of General Medical Sciences or the National Institutes of Health.

References

1. Berman, H. M., Westbrook, J., Feng, Z., Gilliland, G., Bhat, T. N., Weissig, H., Shindyalov, I. N., and Bourne, P. E. (2000) The Protein Data Bank, *Nucleic Acids Res 28*, 235–242.

2. Burley, S. K., Joachimiak, A., Montelione, G. T., and Wilson, I. A. (2008) Contributions to the NIH-NIGMS Protein Structure Initiative from the PSI Production Centers, *Structure 16*, 5–11.

3. Murshudov, G. N., Vagin, A. A., and Dodson, E. J. (1997) Refinement of macromolecular structures by the maximum-likelihood method, *Acta Crystallogr Sect D Biol Crystallogr 53*, 240–255.

4. The CCP4 suite: programs for protein crystallography. (1994) *Acta Crystallogr Sect D Biol Crystallogr 50*, 760–763.

5. Davis, I. W., Leaver-Fay, A., Chen, V. B., Block, J. N., Kapral, G. J., Wang, X., Murray, L. W., Arendall, W. B., 3rd, Snoeyink, J., Richardson, J. S., and Richardson, D. C. (2007) MolProbity: all-atom contacts and structure validation for proteins and nucleic acids, *Nucleic Acids Res 35*, W375–W383.

6. Tickle, I. J., Laskowski, R. A., and Moss, D. S. (1998) Rfree and the Rfree ratio. I. Derivation of expected values of cross-validation residuals used in macromolecular least-squares refinement, *Acta Crystallogr Sect D Biol Crystallogr 54*, 547–557.

7. Tickle, I. J., Laskowski, R. A., and Moss, D. S. (2000) Rfree and the Rfree ratio. II. Calculation of the expected values and variances of cross-validation statistics in macromolecular least-squares refinement, *Acta Crystallogr Sect D Biol Crystallogr 56*, 442–450.

8. Kleywegt, G. J. (1997) Validation of protein models from Calpha coordinates alone, *J Mol Biol 273*, 371–376.

9. Chenna, R., Sugawara, H., Koike, T., Lopez, R., Gibson, T. J., Higgins, D. G., and Thompson, J. D. (2003) Multiple sequence alignment with the Clustal series of programs, *Nucleic Acids Res 31*, 3497–3500.

10. Krissinel, E., and Henrick, K. (2004) Secondary-structure matching (SSM), a new tool for fast protein structure alignment in three dimensions, *Acta Crystallogr Sect D Biol Crystallogr 60*, 2256–2268.

11. Emsley, P., and Cowtan, K. (2004) Coot: model-building tools for molecular graphics, *Acta Crystallogr Sect D Biol Crystallogr 60*, 2126–2132.

12. Feng, Z., Westbrook, J., and Berman, H. M. (1998) NUCheck, NDB-407, Rutgers University, New Brunswick, NJ.

13. Laskowski, R. A., MacArthur, M. W., Moss, D. S., and Thornton, J. M. (1993) PROCHECK: a program to check the stereochemical quality of protein structures, *J Appl Crystallogr 26*, 283–291.

14. Vaguine, A. A., Richelle, J., and Wodak, S. J. (1999) SFCHECK: a unified set of procedures for evaluating the quality of macromolecular structure-factor data and their agreement with the atomic model, *Acta Crystallogr Sect D Biol Crystallogr 55*, 191–205.

15. Terwilliger, T. C. (2000) Maximum-likelihood density modification, *Acta Crystallogr Sect D Biol Crystallogr 56*, 965–972.

16. Krissinel, E., and Henrick, K. (2007) Inference of macromolecular assemblies from crystalline state, *J Mol Biol 372*, 774–797.

17. Henrick, K., and Thornton, J. M. (1998) PQS: a protein quaternary structure file server, *Trends Biochem Sci 23*, 358–361.

18. Ponstingl, H., Kabir, T., and Thornton, J. M. (2003) Automatic inference of protein quaternary structure from crystals, *J Appl Crystallogr 36*, 1116–1122.

19. Camacho, C. J., and Zhang, C. (2005) Fast-Contact: rapid estimate of contact and binding free energies, *Bioinformatics 21*, 2534–2536.

20. Negi, S. S., Schein, C. H., Oezguen, N., Power, T. D., and Braun, W. (2007) InterProSurf: a web server for predicting interacting sites on protein surfaces, *Bioinformatics 23*, 3397–3399.

21. Tina, K. G., Bhadra, R., and Srinivasan, N. (2007) PIC: Protein Interactions Calculator, *Nucleic Acids Res 35*, W473–W476.

22. Jones, S., and Thornton, J. M. (1996) Principles of protein-protein interactions, *Proc Natl Acad Sci USA 93*, 13–20.

23. Saha, R. P., Bahadur, R. P., Pal, A., Mandal, S., and Chakrabarti, P. (2006) ProFace: a server for the analysis of the physicochemical features of protein-protein interfaces, *BMC Struct Biol 6*, 11.

24. Porollo, A., and Meller, J. (2007) Prediction-based fingerprints of protein-protein interactions, *Proteins 66*, 630–645.

25. Han, G. W., Schwarzenbacher, R., Page, R., Jaroszewski, L., Abdubek, P., Ambing, E., Biorac, T., Canaves, J. M., Chiu, H. J., Dai, X., Deacon, A. M., DiDonato, M., Elsliger, M. A., Godzik, A., Grittini, C., Grzechnik, S. K., Hale, J., Hampton, E., Haugen, J., Hornsby, M., Klock, H. E., Koesema, E., Kreusch, A., Kuhn, P., Lesley, S. A., Levin, I., McMullan, D., McPhillips, T. M., Miller, M. D., Morse, A., Moy, K., Nigoghossian, E., Ouyang, J., Paulsen, J., Quijano, K., Reyes, R., Sims, E., Spraggon, G., Stevens, R. C., van den Bedem, H., Velasquez, J., Vincent, J., von Delft, F., Wang, X., West, B., White, A., Wolf, G., Xu, Q., Zagnitko, O., Hodgson, K. O., Wooley, J., and Wilson, I. A. (2005) Crystal structure of an alanine-glyoxylate aminotransferase from Anabaena sp. at 1.70 Å resolution reveals a noncovalently linked PLP cofactor, *Proteins 58*, 971–975.

26. Han, G. W., Schwarzenbacher, R., McMullan, D., Abdubek, P., Ambing, E., Axelrod, H., Biorac, T., Canaves, J. M., Chiu, H. J., Dai, X., Deacon, A. M., DiDonato, M., Elsliger, M. A., Godzik, A., Grittini, C., Grzechnik, S. K., Hale, J., Hampton, E., Haugen, J., Hornsby, M., Jaroszewski, L., Klock, H. E., Koesema, E., Kreusch, A., Kuhn, P., Lesley, S. A., McPhillips, T. M., Miller, M. D., Moy, K., Nigoghossian, E., Paulsen, J., Quijano, K., Reyes, R., Spraggon, G., Stevens, R. C., van den Bedem, H., Velasquez, J., Vincent, J., White, A., Wolf, G., Xu, Q., Hodgson, K. O., Wooley, J., and Wilson, I. A. (2005) Crystal structure of an Apo mRNA decapping enzyme (DcpS) from Mouse at 1.83 Å resolution, *Proteins 60*, 797–802.

27. Mathews, I. I, McMullan, D., Miller, M. D., Canaves, J. M., Elsliger, M. A., Floyd, R., Grzechnik, S. K., Jaroszewski, L., Klock, H. E., Koesema, E., Kovarik, J. S., Kreusch, A., Kuhn, P., McPhillips, T. M., Morse, A. T., Quijano, K., Rife, C. L., Schwarzenbacher, R., Spraggon, G., Stevens, R. C., van den Bedem, H., Weekes, D., Wolf, G., Hodgson, K. O., Wooley, J., Deacon, A. M., Godzik, A., Lesley, S. A., and Wilson, I. A. (2007) Crystal structure of 2-keto-3-deoxygluconate kinase (TM0067) from *Thermotoga maritima* at 2.05 Å resolution, *Proteins 70*, 603–608.

28. Kim, S. H., Shin, D. H., Choi, I. G., Schulze-Gahmen, U., Chen, S., and Kim, R. (2003) Structure-based functional inference in structural genomics, *J Struct Funct Genomics 4*, 129–135.

29. Holm, L., and Sander, C. (1996) Alignment of three-dimensional protein structures: network server for database searching, *Methods Enzymol 266*, 653–662.

30. Martin, A. C. (2000) The ups and downs of protein topology; rapid comparison of protein structure, *Protein Eng 13*, 829–837.

31. Gibrat, J. F., Madej, T., and Bryant, S. H. (1996) Surprising similarities in structure comparison, *Curr Opin Struct Biol 6*, 377–385.

32. Crowley, C., Sawaya, M. R., and Yeates, T. O. Crystal structure of PduU from the ethanolamine microcompartment of Salmonella, *In Preparation*.

33. Kerfeld, C. A., Sawaya, M. R., Tanaka, S., Nguyen, C. V., Phillips, M., Beeby, M., and Yeates, T. O. (2005) Protein structures forming the shell of primitive bacterial organelles, *Science 309*, 936–938.

34. Kawabata, T. (2003) MATRAS: a program for protein 3D structure comparison, *Nucleic Acids Res 31*, 3367–3369.

35. Ye, Y., and Godzik, A. (2004) FATCAT: a web server for flexible structure comparison and structure similarity searching, *Nucleic Acids Res 32*, W582–W585.

36. Shindyalov, I. N., and Bourne, P. E. (1998) Protein structure alignment by incremental combinatorial extension (CE) of the optimal path, *Protein Eng 11*, 739–747.

37. Gaspari, Z., Vlahovicek, K., and Pongor, S. (2005) Efficient recognition of folds in protein

3D structures by the improved PRIDE algorithm, *Bioinformatics 21*, 3322–3323.

38. Shyu, C. R., Chi, P. H., Scott, G., and Xu, D. (2004) ProteinDBS: a real-time retrieval system for protein structure comparison, *Nucleic Acids Res 32*, W572–W575.

39. Pearl, F. M., Bennett, C. F., Bray, J. E., Harrison, A. P., Martin, N., Shepherd, A., Sillitoe, I., Thornton, J., and Orengo, C. A. (2003) The CATH database: an extended protein family resource for structural and functional genomics, *Nucleic Acids Res 31*, 452–455.

Chapter 9

Knowledge Discovery via Machine Learning for Neurodegenerative Disease Researchers

I. Burak Özyurt and Gregory G. Brown

Summary

Ever-increasing size of the biomedical literature makes more precise information retrieval and tapping into implicit knowledge in scientific literature a necessity. In this chapter, first, three new variants of the expectation–maximization (EM) method for semisupervised document classification (Machine Learning 39:103–134, 2000) are introduced to refine biomedical literature meta-searches. The retrieval performance of a multi-mixture per class EM variant with Agglomerative Information Bottleneck clustering (Slonim and Tishby (1999) Agglomerative information bottleneck. In *Proceedings of NIPS-12*) using Davies–Bouldin cluster validity index (IEEE Transactions on Pattern Analysis and Machine Intelligence 1:224–227, 1979), rivaled the state-of-the-art transductive support vector machines (TSVM) (Joachims (1999) Transductive inference for text classification using support vector machines. In *Proceedings of the International Conference on Machine Learning (ICML)*). Moreover, the multi-mixture per class EM variant refined search results more quickly with more than one order of magnitude improvement in execution time compared with TSVM. A second tool, CRFNER, uses conditional random fields (Lafferty et al. (2001) Conditional random fields: Probabilistic models for segmenting and labeling sequence data. In *Proceedings of ICML-2001*) to recognize 15 types of named entities from schizophrenia abstracts outperforming ABNER (Settles (2004) Biomedical named entity recognition using conditional random fields and rich feature sets. In *Proceedings of COLING 2004 International Joint Workshop on Natural Language Processing in Biomedicine and its Applications (NLPBA)*) in biological named entity recognition and reaching F_1 performance of 82.5% on the second set of named entities.

Key words: Relevance ranking, Information extraction, EM, Cluster validity, Conditional random fields, Machine learning, Knowledge discovery, Schizophrenia

---

1. Introduction

Most human knowledge is stored in written form encoded in natural languages easily decodable by humans but too unstructured for computers. While keyword-based search engines ease a

Vadim Astakhov (ed.), *Biomedical Informatics,* Methods in Molecular Biology, vol. 569
DOI 10.1007/978-1-59745 524 4_9, © Humana Press, a part of Springer Science+Business Media, LLC 2009

researcher's information sifting task, they are far from being perfect. Both general-purpose and domain-specific search engines suffer from low precision rate. Lack of search precision requires users to search through retrieved results to identify relevant documents. In addition, encoded knowledge in scientific documents is far richer than keywords and has complex relational structure and vastly diverse representational form evading discovery by keywords. Thus, extracting structured information for building machine-queryable knowledge bases from unstructured textual data is of increasing importance to aid researchers' quest for knowledge in ever-growing biomedical literature. The premier bibliographic database of the National Library of Medicine, MEDLINE (*see* **Note 1**), for example, grows at a rate of more than a half million articles every year.

In this chapter, two complementary tools for aiding knowledge discovery in neurodegenerative disease research are presented. The first tool is a search result refinement system that is a part of the meta-search engine for the Biomedical Informatics Research Network (BIRN) (http://www.nbirn.net) Query Atlas *(7)*. Query Atlas combines browsing/analysis of functional and structural magnetic resonance imaging ((f)MRI) data with text/literature mining. Three new extensions to the expectation maximization (EM) algorithm *(1)* for text classification from labeled and unlabeled data are introduced and compared against transductive Support Vector Machines (SVMs) *(4)*. The second tool described in **Subheading 3** uses conditional random fields (CRFs) *(5)* to recognize named entities in schizophrenia abstracts. Named entities like age and dosage information, protein, drug, and disease names are not only useful by themselves in building structured databases and/or controlled vocabularies, they are also useful as building blocks and/or features of higher-order NLP systems that identify and exploit relations between named entities and other language constructs.

2. Search Result Refinement via Semisupervised Machine Learning for Biomedical Journal Search

Most search engines use term-based queries with different document index model representations and document ranking functions. To do a search query, the user must translate the context of his/her intent into search terms. Usually this translation is at best partial; hence the search results reflect this partial context. Provided with search results, however, the user can indicate a small set of results as relevant providing feedback to the information retrieval system to rerank the search results for higher precision rate.

Relevance feedback, one of the most popular query reformulation strategies for information retrieval (IR), is based on query expansion and/or term reweighting techniques available for vector and probabilistic models of information retrieval *(8)*. The semisupervised machine learning approach to relevance feedback taken here is akin to the Bayesian classification model of retrieval *(9)*, where the relevant documents are used to model the relevant class for the query and the remaining corpus for the nonrelevant class. The main difference is the incorporation of unlabeled data for better estimation of both relevant and nonrelevant classes in the classification model.

As in *(10)*, multiple topics per class EM extensions use a language modeling perspective. Unlike *(10)*, however, where the language model is applied to the user query, the introduced extensions represent the relatively more abundant nonrelevant labeled documents as a mixture of unknown but to be estimated number of topics.

2.1. Transductive SVMs

In natural languages, words occur in strong co-occurrence patterns *(11)*. Some words are more likely to occur in related documents than the unrelated documents. This phenomenon is exploited by the unsupervised document categorization approaches. Transductive SVMs *(4)* exploit this phenomenon to improve the classification performance over limited number of labeled training data by incorporating the readily available unlabeled data also in learning.

2.2. Mixture of Unigrams Expectation Maximization from Labeled and Unlabeled Data

In statistical terms, a document d_i from a corpus $D = \{d_i\} i = 1, \ldots, N$ can be seen as generated by first selecting a topic from the set of topics $C = \{c_j\} j = 1, \ldots, J$, then selecting a document length $|d_i|$, and finally selecting $|d_i|$ words sequentially from a vocabulary $V = \{w_t\}$. In order to accurately model the particular order of words making up a document, strong order dependencies in human languages must be taken into account, resulting in intractable chain of conditional probabilities. To make the model tractable, usually word independence assumption, aka Naive Bayes or bag of words assumption, is used. While modeling bias with Naive Bayes is high, under zero-one loss function for classification error, the Naive Bayes document model can work very well for classification, where the only requirement is a negative boundary bias *(12)*. A corpus using Naive Bayes assumption, thus, can be modeled as a mixture of topic components, where each document is allowed to belong to a single topic component, parametrized by the parameter set $\theta = \theta_{w_t|c_j} : w_t \in V, c_j \in C; \theta_{c_j} \in C$. The probability that document d_i given the topic c_j then can be expressed as

$$p(d_i \mid c_j; \hat{\theta}) = p(\mid d_i \mid) \prod_{k=1}^{|d_i|} p(w_{t,d_{i,k}} \mid c_j; \hat{\theta}) \tag{1}$$

By maximum a posteriori (MAP) point estimate, i.e., $\text{argmax}_\theta p(\theta|D)$, using Dirichlet prior, conjugate prior of multinomial distribution, with $\alpha = 2$, we end up with a Naive Bayes classifier with Laplace smoothing to prevent overfitting.

The word probability estimates $\theta_{w_t|c_j} \equiv p(w_t \mid c_j; \hat{\theta})$ are expressed as

$$\hat{\theta}_{w_t|c_j} = \frac{1 + \sum_{i=1}^{|D|} N(w_t, d_i) p(y_i = c_j \mid d_i)}{|V| + \sum_{s=1}^{|V|} \sum_{i=1}^{|D|} N(w_s, d_i) p(y_i = c_j \mid d_i)} \tag{2}$$

The class prior probabilities with Laplace smoothing are

$$p(y_i \mid d_i; \hat{\theta}) = \frac{1 + \sum_{i=1}^{|D|} p(y_i = c_j \mid d_i)}{|C| + |D|} \tag{3}$$

The Naive Bayes classifier then uses the parameters calculated from training documents to estimate the most likely class y_i for a new document by using Bayes' rule:

$$p(y_i = c_j \mid d_i; \hat{\theta}) = \frac{p(c_j \mid \hat{\theta}) \prod_{k=1}^{|d_i|} p(w_{t,d_{i,k}} \mid c_j; \hat{\theta})}{\sum_{r=1}^{|C|} p(c_r \mid \hat{\theta}) \prod_{k=1}^{|d_i|} p(w_{t,d_{i,k}} \mid c_r; \hat{\theta})} \tag{4}$$

For Nigam and colleagues' *(1)* EM approach, the training dataset consists of a small set of labeled documents and a much larger set of unlabeled documents $D = D^l \cap D^u$, which may convey further information in the form of co-occurrence of words, both found in labeled and unlabeled documents.

In standard EM for mixture components, the unobserved (latent) mixing proportions (**Z**) are estimated from the observed data (X) by finding the parameters $\hat{\theta}$ that maximize the log likelihood of complete data $L(\mathbf{X},\mathbf{Z}|\theta)$.

Basic EM method for one mixture component per class starts with a "priming" M-step, i.e., estimating parameters for the Naive Bayes classifier from just the labeled set by **Eqs. 2** and **3**. The EM iterations, then, begin with an E-step using the Naive Bayes classifier **Eq. 4** to estimate the most likely class/mixture component for the unlabeled documents followed by the M-step where the new MAP estimates for parameters $\hat{\theta}$ are calculated using the current estimates for $p(c_j \mid d_i; \hat{\theta})$ and by **Eqs. 2** and **3**.

2.2.1. Multiple Mixture Component/Topic per Class Case

It is restrictive to assume that for each class documents will belong to a single topic. Nigam et al. *(1)* extended their basic EM method to multi topic/mixture component per class model. Using c_j to continue to denote the *j*th mixture component, we can define the *a*th class as G_a. Then, the class probability of a document can be expressed as the weighted sum of the mixture component probabilities:

$$p(G_a \mid d_i; \hat{\theta}) = \sum p(G_a \mid c_j; \hat{\theta}) \frac{p(c_j \mid \hat{\theta}) \prod_{k=1}^{|d_i|} p(w_{t,d_{i,k}} \mid c_j; \hat{\theta})}{\sum_{r=1}^{|C|} p(c_r \mid \hat{\theta}) \prod_{k=1}^{d_i} p(w_{t,d_{i,k}} \mid c_r; \hat{\theta})} \quad (5)$$

2.3. Multiple Mixture per Class Component Number and Member Selection by Hierarchical Clustering and Cluster Validity

Nigam et al. *(1)* report using cross-validation for mixture component/topic number selection and they uniformly distribute the labeled documents between the selected numbers of mixture components. However, in an online search refinement situation, what we can expect from the user is to indicate at most a few documents as relevant documents. The negative class consists of the documents in between the relevant ones and usually much larger than the positive (relevant) class. However, the most important constraint is the execution time. The user expects the filtered results within a few seconds, unlike offline document classification. The disadvantages of cross-validation in this situation are both the scarcity of the training data and most importantly time needed to do cross-validation which can be prohibiting. Due to the stochastic nature of mixture component initialization, in determining the number of mixture components multiple random initializations at each tested mixture component count will also be necessary; thus multi-folding the cross-validation time.

The effect of random initializations at different mixture component/cluster numbers ranging from 2 to 10 for the negative class for Nigam and colleagues' multiple mixture components per class EM (MMEM) for some of the datasets with most variance is shown in **Fig. 1**. For the nine test cases used in this study, eight had enough data to do this analysis. For each of these eight datasets, for each cluster number from 2 to 10, 30 random initializations for MMEM are done. As the results show, MMEM is rather sensitive both to the number of mixture components and to the initialization, which demonstrates the need for multiple initializations at multiple cluster numbers for cross-validation. For each dataset 30 initializations for 2–10 clusters took roughly 10 min on a Pentium 4 3.0 GHz machine.

In the light of these findings a more time efficient approach for simultaneous mixture component initialization and "optimal" cluster number determination by using an agglomerative hierarchical clustering algorithm *(13)* with cluster validity based

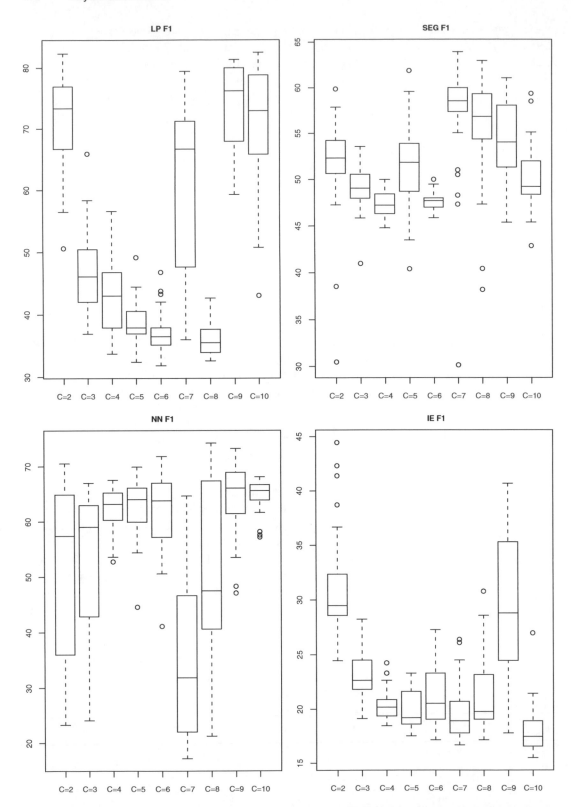

Fig. 1. Test F_1 percentage values for 30 random mixture component initializations for negative (non-relevant) class at cluster numbers 2–10.

cluster number determination is introduced. Hierarchical clustering has the advantages of being applicable to cases where we are unable to supply a distance metric, but pairwise dissimilarity values for every pair of samples and generating a cluster hierarchy as the output. Symmetric Kullback–Leibler (KL) divergence is used as the dissimilarity value between pairs of search result documents each comprising a title and (if available) an abstract. Symmetric KL is defined as

$$sD_{KL}[p(x) \| q(x)] = 0.5(D_{KL}[p(x) \| q(y)] + D_{KL}[q(y) \| p(x)] \quad (6)$$

$$D_{KL}[p(x) \| q(x)] = \sum_x p(x) \log \frac{p(x)}{q(x)}$$

A Farthest-Neighbor Agglomerative clustering algorithm based on symmetric KL divergence as dissimilarity measure between document word probability distributions is shown in Algorithm 1.

Although hierarchical clustering approaches have $O(cn^2d^2)$ time complexity and $O(n^2)$ space complexity and potentially result in suboptimal clustering due to the greedy nature of the algorithms and nonconvexity of most clustering problems, these do not pose a problem for online result refinement, since the training dataset (i.e., n) is rather small due to the nature of the task, hence both time and space complexity are manageable. Here, n is the number of documents, d is the dimension of the solution space, and c is the number of clusters. Suboptimal clustering due to local minima is the common problem of all clustering algorithms, for which there is no time efficient solution.

Many cluster validity indices are introduced over time to evaluate partitioning induced by clustering algorithms *(14, 3, 15)*. Based on Maulik and Bandyopadhyay's *(15)* comparison study

Algorithm 1
Farthest-Neighbor Agglomerative Hierarchical Clustering

Initialize $\hat{c} \Leftarrow n, C_i \Leftarrow \{d_i\} i = 1, \ldots, n$

while $\hat{c} > 1$ **do**

    find nearest clusters by

$$\arg_{i, j} \max_{p(x) \in C_i, q(x) \in C_j} sD_{KL}[p(x) \| q(x)]$$

    merge clusters C_i and C_j

    $\hat{c} \Leftarrow \hat{c} - 1$

end while

return cluster hierarchy

of the most common cluster validity indices on center-based clustering algorithms, the Davies–Bouldin index *(3)* is selected to determine the "optimal" cluster number and adapted for pairwise clustering. This index is a function of the ratio of the sum of within-cluster scatter to between-cluster separation. The "optimal" cluster number occurs at the minimum of the DB index with increasing number of clusters. In practice, sometimes cluster validity indices keep decreasing with the increase of the number of clusters. In this case, the minimum slope of the DB index curve can be used as an indicator where the addition of another cluster will have only a marginal effect.

The within-cluster scatter for the ith cluster is calculated as

$$S_i = \frac{1}{|C_i|} \sum_{p_j(x) \in C_i} sD_{\mathrm{KL}}[p_j(x) \| q(x)]$$

(7)

where $q(x)$ is the word probability distribution for the centroid. The centroid of a cluster is defined as the document which has the least maximum dissimilarity from the other cluster documents, i.e.,

$$\arg \min_j \max_{k, k \neq j} sD_{\mathrm{KL}}[p_j(x) \| p_k(x)]$$

(8)

Cluster separation d_{ij} between cluster C_i and C_j is defined as $sD_{\mathrm{KL}}[q_i(x) \| q_j(x)]$, where $q_i(x)$ is the word probability distribution for the centroid of cluster i. The Davies–Bouldin (DB) index is defined as

$$DB = \frac{1}{K} \sum_{i=1}^{k} \max_{j, j \neq 1} \left\{ \frac{S_i + S_j}{d_{ij}} \right\}$$

(9)

The "optimal" number of clusters is determined by the minimum or the first zero crossing of the negative of the scaled numerical derivative of the DB(t) function

$$-\frac{DB_t - DB_{t+1}}{DB_t}$$

where t, corresponding to K in **Eq. 9**, denotes the number of clusters for a particular partition.

2.3.1. Multiple Mixture per Class Component Number/ Member Selection by Agglomerative Information Bottleneck and Cluster Validity

Agglomerative Information Bottleneck (AIB) *(2)* is a hierarchical, bottom-up, distributional hard clustering algorithm that maximizes the mutual information per cluster between the data and given categories. AIB, a variant of Information Bottleneck method, finds a compressed version of variable X, corresponding to the clusters Z, such that the mutual information between Z

and a relevant variable Υ, $I(Z,\Upsilon)$, which is conditionally dependent on X but independent of Z (i.e., forming a Markov chain $Z \rightarrow X \rightarrow \Upsilon$), is maximized under a constraint on the mutual information between X and Z. The solution to this constrained optimization problem results in self-consistent equations for the conditional distributions $p(y|z)$, $p(x|z)$, and $p(z)$. AIB uses Jensen–Shannon divergence as the distortion measure between conditional distributions of the relevant variable given clusters $p(y\,|\,z_i),: i \in \{1,...,M\}$,

$$JS_p[p(y\,|\,z_1),...,p(y\,|\,z_M)]$$
$$= \sum_{i=1}^{M} \pi_i D_{KL}[p(y\,|\,z_i)\,||\,\sum_{i=1}^{M} \pi_i p(y,z_i)] \qquad (10)$$

Here π_i is the prior probability of z_i. For document classification, X corresponds to documents, Υ corresponds to terms occurring in a document, and Z corresponds to the document clusters we are after.

Davies–Bouldin index (3) is adapted for AIB using Jensen–Shannon (JS) divergence instead of symmetric KL divergence in a similar fashion as defined before. The JS divergence is calculated between each pair of documents in each cluster for Davies–Bouldin index calculations. Also, the "optimal" number of clusters is determined in a similar fashion to the farthest-neighbor hierarchical clustering algorithm.

2.4. Simulated Annealing EM for Labeled/ Unlabeled Document Classification

Simulated annealing (16) is a well-known stochastic search technique used to find global optimum in problems with multiple local extrema. Simulated annealing has a temperature parameter similar to its physical counterpart. Valleys and peaks of the solution space with objective function value differences less than the temperature value become the search space for simulated annealing at that temperature. The annealing process starts at a high temperature, where the data points can belong to any mixture more or less with equal probability; the temperature is gradually lowered according to a cooling scheme, at zero temperature a data point will belong to a single mixture component with probability 1. For simulated annealing EM, the E-step of EM as defined by **Eq. 4** is replaced by

$$p(y_i = c_j \,|\, d_i; \hat{\theta}) = \frac{p(c_j \,|\, \hat{\theta}) \exp\left\{\dfrac{\log p(d_i \,|\, c_j; \hat{\theta})}{T}\right\}}{\sum_{r=1}^{|C|} p(c_r \,|\, \hat{\theta}) \exp\left\{\dfrac{\log p(d_i \,|\, c_r; \hat{\theta})}{T}\right\}} \qquad (11)$$

where T is the temperature parameter and $p(d_i \mid c_j; \hat{\theta})$ values are calculated by **Eq. 1**. The derivation for this method is given in *(17)*.

2.5. Experimental Design

To test the classification methods described above, the introduced meta-search engine is used to query the National Library of Medicine's search service PubMed (*see* **Note 2**). A typical query result consists of the title, authors, journal information, and abstract of a paper. The Query Atlas meta-search engine (*see* **Note 3**) returns by default 200 top-ranking search results from PubMed. Nine queries on topics the first author has some expertise about are used to generate test datasets. The queries are shown in **Table 1**. From this query result set, the ones which seem relevant to the author are tagged as relevant. For example, for "neural network" query the relevant documents are related to machine learning algorithms but not to the neural networks in the brain. The goal of an efficient search refinement algorithm is to detect complex implicit dependency patterns in relevant documents, which cannot be expressed explicitly in search queries. Overall 1,800 abstracts are tagged as either relevant or nonrelevant. For training, the first few relevant documents scanning from the top-ranked document down are used as the positive examples and the nonrelevant examples up to the last positive example make up the negative examples. The number of positive examples is selected such that the negative class has enough data to test mixture component clustering and EM algorithm improved Naive Bayes.

Table 1
Datasets

Document set	# Relevant (training)	# Irrelevant (training)	# Relevant (test)
Language processing (LP)	2	25	29
Segmentation (SEG)	4	12	50
Neural networks (NN)	4	21	82
Information extraction (IE)	4	53	12
Text categorization (TC)	4	4	41
Ontology (Ont)	9	17	83
Active contour (AC)	9	22	31
Support vector machines (SVM)	5	13	35
Semantic (SEM)	4	30	32

The remaining documents in each 200-document dataset constitute the test cases. The hand-tagged document sets are summarized in **Table 1**. Standard information retrieval performance measures *precision P*, *recall R*, and harmonic mean of precision and recall F_1 are used. These measures are defined as

$$P = \frac{|\{\text{relevant docs}\} \cap \{\text{retrieved docs}\}|}{|\{\text{retrieved docs}\}|} = \frac{\text{TP}}{\text{TP} + \text{FP}} \quad (12)$$

$$R = \frac{|\{\text{relevant docs}\} \cap \{\text{retrieved docs}\}|}{|\{\text{relevant docs}\}|} = \frac{\text{TP}}{\text{TP} + \text{FN}} \quad (13)$$

$$F_1 = \frac{2PR}{P + R} \quad (14)$$

Here TP, FP, and FN stand for true positive, false positive, and false negative counts, respectively. Only the title, abstract, and journal title of a search result are used for classification. The numeric values, special characters and stop-words in the documents are discarded. The common tf-idf term weighting $N(w_t, d_i) \log(|D| / |d_i \supset w_t|)$ for word/term frequencies is used in all experiments instead of pure word frequencies. For KL and JS divergence calculations, Laplace smoothing is used to prevent zero probability problems. For simulated annealing version of EM, the starting temperature T_0 used is 5. Temperature is decreased linearly down to 1.0 in ten iterations of EM. All EM variants are run for ten iterations.

2.6. Experimental Results and Discussion

The results for Naive Bayes (NB), EM with single mixture component per class (EM), MMEM with symmetric KL divergence-based hierarchical clustering with Davies–Bouldin (DB) cluster validity index (MMEM(KL)), EM with Agglomerative Information Bottleneck clustering with DB cluster validity index (MMEM(AIB)), simulated annealing EM with single mixture component per class (SAEM), K-nearest neighbor classifier with $k = 1$ (KNN), and Transductive SVM (TSVM) are summarized in **Tables 2** and **3**.

Based on the average F_1 values, MMEM(AIB) performed best followed by TSVM and SAEM. SAEM and MMEM(AIB) were also about 60 times and 20 times faster than TSVM on average, respectively. Single mixture component per class EM for the datasets SEG, NN, and SEM showed the worst performance in all the EM methods leaving room for improvement. For SEG, SAEM was able to find a much better local maximum, but did only slightly better for NN and SEM datasets. Overall, all of the newly introduced EM variants achieved better average F_1 values than Nigam and colleagues' EM. For dataset SVM, single

Table 2
F_1 percentages for tested methods

	LP	SEG	NN	IE	TC	Ont	AC	SVM	SEM
NB	45.0	10.5	6.7	28.6	62.3	63.5	51.2	40.0	10.5
EM	70.8	22.6	21.8	64.5	63.4	74.5	48.6	62.5	14.3
MMEM_KL	85.2	24.6	18.6	66.7	63.4	75.3	48.1	56.1	18.6
MMEM_AIB	74.6	46.5	51.5	45.8	63.4	74.9	48.6	62.3	69.8
SAEM	72.0	60.0	25.5	68.8	64.0	81.4	43.3	63.8	18.2
KNN	61.2	40.0	34.2	34.5	55.4	64.7	41.3	47.4	39.2
TSVM	61.9	54.2	39.6	54.5	63.0	71.3	57.5	55.8	61.5

Table 3
Results summary for F_1 percentages and average execution times in seconds

	NB	EM	MMEM_KL	MMEM_AIB	SAEM	KNN	TSVM
Avg F_1	35.4	49.2	50.7	59.7	55.2	46.4	57.7
Time	0.7	1.1	3.5	3.6	1.2	1.0	72.4
SD	0.2	0.3	2.9	2.8	0.3	0.3	60.6

mixture component per class EM had better performance than MMEM variants. The determination of the degenerate case single "optimal" cluster is not possible with cluster validity indices, which are not even defined for a single cluster. Levine and Domany *(18)* proposed a cluster validity method based on resampling, which can detect if the data has no cluster tendency. However, the data size requirements and time complexity of this method make it infeasible for search results refinement. Also, for dataset AC, EM and all its variants deteriorated Naive Bayes performance. Similar behavior is also reported in *(1)*. SAEM performance for dataset AC was worse than EM. This was due to the fact that the cooling scheme is suboptimal. It is shown in *(19)* that in theory global optimum can be achieved if the cooling schedule obeys $T \propto 1 / \log n$ where n is the number of current iteration, which is unrealistic even for offline applications. While the performance improvement over TSVM for MMEM(AIB) is not statistically significant using paired Wilcoxon signed rank test at $\alpha = 0.05$, the improvement in execution time, which is essential for the applicability of a method in search results refinement, is statistically significant.

From the random initialization and different mixture component number experiments before, it seems that multi-mixture component EM is sensitive to initialization and number of mixture components. Both clustering-based approaches introduced provide a one-pass, automatic way of selecting the number of mixture components and initialization of them while providing especially for MMEM(AIB), substantial improvement over one mixture component per class EM within user acceptable execution time. SAEM, on the other hand, assumes one mixture component per class, and tries to avoid getting stuck in a local maximum while still operating within user acceptable execution times.

3. Information Extraction from PubMed Abstracts via Conditional Random Fields

In this section, conditional random fields (CRFs) are applied to a particular natural language processing (NLP) task, namely, named entity recognition (NER), to detect 15 named entity types from schizophrenia abstracts. CRFs are successfully applied to recognize titles, abstracts, authors, keywords, etc. in computer science papers *(20)* and to recognize biological named entities in biomedical abstracts *(6)*. Named entities like dosage information, protein, drug, and disease names are not only useful by themselves in building databases and/or controlled vocabularies, they are also useful for higher-order NLP tasks including semantic role labeling and question answering. Here, we present a CRF-based named entity recognizer, CRFNER, extended with syntactic and semantic features that outperforms ABNER *(6)* on the schizophrenia corpus for biological NER and allows for recognition of ten additional named entities deemed to be important in building a structured knowledge base for neurodegenerative disease knowledge discovery and question answering.

3.1. Conditional Random Fields

A classification task of predicting outputs **Y** from provided inputs **X** can be approached, probabilistically, by estimating the conditional probability $P(\mathbf{Y}|\mathbf{X})$. A discriminative classifier models this conditional probability directly, whereas a generative classifier first models the joint probability $P(\mathbf{X},\mathbf{Y})$. Modeling of the joint probability $P(\mathbf{X},\mathbf{Y})$ involves taking into account intricate interactions between input variables, which can easily make the parameter estimation task intractable. However, a discriminative classifier avoids this problem by directly estimating the conditional probability, while allowing complicated input interdependencies to be expressed.

A CRF *(5)* is a Markov random field that is globally conditioned on input **X**. A Markov random field *(21)* is a graphical model based on an undirected graph $G{:}(V,E)$, where each vertex $v \in V$ corresponds to a random variable and the joint probability distribution $p(v_1,\ldots,v_n)$ obeys the local Markov property, i.e.,

$$\forall v \in V \; p(v \mid V \setminus \{v\}) = p(v \mid n(v))$$

Here $n(v)$ denotes the set of neighbors of vertex v.

By the fundamental theorem of random fields *(22)*, the parametrized conditional probability $p_\Lambda(\mathbf{y} \mid \mathbf{x})$ of a Markov random field can be expressed as

$$p_\Lambda(\mathbf{y} \mid \mathbf{x}) = \frac{1}{Z(x)} \prod_{c \in C(x,y)} \Phi(\mathbf{x}_c, \mathbf{y}_c)$$

Here, $C(\mathbf{x},\mathbf{y})$ is the set of cliques of the graph $G, \Phi(\mathbf{x}_c, \mathbf{y}_c)$ is a nonnegative clique potential function, and $Z(\mathbf{x})$ is partition function normalizing the product of clique potentials defined as

$$Z(x) = \sum_{\mathbf{y}'} \prod_{c \in C(\mathbf{x},\mathbf{y}')} \Phi\left(\mathbf{x}_c, \mathbf{y}'_c\right)$$

For the NER task, an input data instance corresponds to the sequence of the words of a sentence from an abstract $\{x_t\} t = 1,\ldots,T$ and the corresponding output sequence are labels in (extended) CoNLL IOB format $\{y_t\} t = 1,\ldots,T$. These labels indicate if their corresponding sentence word is a part of a named entity or not. A tagged sentence example is shown in **Fig. 2**.

One particular type of CRF model that is particularly suitable for modeling natural language sequences is linear-chain CRF, which can be considered forming a discriminative–generative pair with hidden Markov models (HMM) *(23)*. Unlike HMMs and Maximum Entropy Markov models, CRFs do not suffer from label bias problem *(5)*. A first-order linear-chain CRF is defined as

$$p_\Lambda(\mathbf{y} \mid \mathbf{x}) = \frac{1}{Z(\mathbf{x})} \exp\left(\sum_{t=1}^{T} \sum_{k=1}^{K} \lambda_k f_k(\mathbf{y}_{t-1}, \mathbf{y}_t, \mathbf{x}, t) \right)$$

In this log-linear model, $f_k(\mathbf{y}_{t-1}, \mathbf{y}_t, \mathbf{x}, \mathbf{t})$ is one of the k feature functions depending potentially on all input variables and only on the current and previous output values, and $\Lambda = \{\lambda_k\}$ is the set of weights to be estimated.

Parameter estimation is done by regularized maximum likelihood or, in Bayesian statistical terms, by MAP estimation, for this particular case, using a Gaussian prior. Thus, the goal is to find parameter Λ values maximizing $p(\mathbf{x},\mathbf{y}) = \mathbf{p}(\mathbf{y}|\mathbf{x})\mathbf{p}(\mathbf{x})$. By monotonic log transformation, penalized likelihood becomes

```
Expression|O:{NN,NPC} of|O:{IN} glutamate|B-PROTEIN:{NN,NPC}
carboxypeptidase|I-PROTEIN:{VBP} II|I-PROTEIN:{NNP,NPC}
in|O:{IN} human|O:{JJ} brain|O:{NN,NPC} .|O:{.}
```

Fig. 2. An example sentence labeled via extended IOB format. The prefix B denotes the beginning on a named entity. The rest of the terms in the NE are denoted by the prefix I and terms not belonging to any NE by O. IOB format is extended to include additional information, namely, part-of-speech tags and noun phrase membership.

$$l(\Lambda) = \log p(\mathbf{y} \mid \mathbf{x}; \Lambda) + \log(\mathbf{x}; \Lambda)$$

Using a Gaussian prior with zero mean and covariance $I\sigma^2$, i.e.,

$$p(\mathbf{x}; \Lambda) \propto \prod_{k=1}^{K} \frac{1}{\sigma\sqrt{2\pi}} \exp\left(\frac{-\lambda_k^2}{2\sigma^2}\right)$$

which penalizes large weights preventing overfitting of the model:

$$l(\Lambda) = \sum_{i=1}^{N}\sum_{t=1}^{T}\sum_{k=1}^{K} \lambda_k f_k\left(\mathbf{y}_{t-1}^{(i)}, \mathbf{y}_t^{(i)}\right) - \sum_{i=1}^{N} \log Z(\mathbf{x}^{(i)}) - \sum_{k=1}^{K} \frac{\lambda_k^2}{2\sigma^2} + C$$

The log likelihood $l(\Lambda)$ is concave and hence has a global maximum. Any gradient base optimization method can be used to maximize it. However, in practice, optimization methods with first-order convergence characteristics usually are very slow to converge. Methods with second-order convergence characteristics, like Newton's method, need calculation and storage of Hessian, whose size is quadratic in number of parameters, which can range for NLP applications in millions, making then impractical. Limited memory versions of quasi-Newton optimization methods like L-BFGS *(24)* were successfully applied for CRF parameter estimation.

An unseen sentence is labeled using the linear-chain CRF-adapted version of HMM's Viterbi algorithm. The Partition function values required during parameter estimation are calculated efficiently by CRF-adapted version of HMM's dynamic programming-based forward–backward formulas.

3.2. Datasets and Preprocessing

To proceed with the goal of creating a structured knowledge database for neurodegenerative disease researchers from unstructured textual data, the first 1,000 abstracts returned from a PubMed search for the keyword "schizophrenia" are selected as the unstructured corpus. This dataset consists of abstract title and body; author and journal information is not used. Each abstract body is first separated into individual sentences by a sentence boundary detector. The implemented sentence boundary detector takes into account acronyms, decimal numbers, etc, since they can be easily mistaken for sentence endings, resulting in spurious sentences. The detected sentences are parsed using Charniak's syntactic parser *(25)*, which also provides part-of-speech (POS) tags for the parsed sentences.

There are two sets of named entities of interest, the first being biological named entities, namely, Protein, DNA, RNA, Cell Line, and Cell Type; the second being combination of generic named entities consisting of Time, Location, Organization, Nationality, and Percentage, and named entities more specific to FBIRN objectives, namely, Drug, Disease, Dosage, Age, and Clinical Assessment. In total, 15 named entities have to be recognized. For the biological named entity set, ABNER is used to bootstrap the hand labeling of 8,800 sentences for the 1,000 PubMed abstracts. The hand labeling is performed by a biochemist/chemical engineer specialized in molecular biology. For the second named entity set, handcrafted regular expressions are used to select a subset of sentences for hand tagging and bootstrapping. In this way, 3,662 sentences are selected and hand-labeled by the first author.

3.3. Features

For the NER task, two sets of binary features are used. The first set consists of mostly orthographic features commonly used in other NER systems and approaches, including in the identity of the word at time t in the sequence (sentence), if the word is all in uppercase, has a certain prefix, etc. The second set of features, here termed the extended feature set, include syntactic features (e.g., POS tags, word is part of a nonrecursive noun phrase) and semantic ones (e.g., word/phrase is in the list of word regions or word has (not necessarily immediate) hyponym of certain more general semantic category in a lexico-semantic database). These two sets of features are listed in **Table 4**. The feature set is further enhanced by conjunctions of orthographic features for the word at current time step in the sequence with the features of words at previous and next time steps in the sequence.

Country and region lists are extracted from the online version of CIA World Factbook. The non-recursive noun phrases are tagged by using a transformation learning-based noun phrase chunker *(26)*. As in ABNER *(6)*, words are also assigned to generalized word classes as used in *(27)*.

As the semantic world knowledge source, WordNet *(28)* version 3.0, a large computerized lexical database of English, is used. WordNet groups English nouns, verbs, adjectives, and adverbs by similarity into synonym sets, which are interlinked via both semantic and lexical relations. As a semantic feature, we used the hypernymy–hyponymy semantic relation for nouns and noun

Table 4
Features used in CRFNER

Feature name	Description
Baseline features	
INITCAPS	Starts with a capitalized letter
INITCAPSALPHA	All-letter word starting with a capitalized letter
ALLCAPS	All characters uppercase
MIXCAPS	Mixture of upper- and lowercase letters
HASDIGIT	At least one character is a digit
NATURALNUMBER	Word is a natural number
REALNUMBER	Word is a decimal number
HASDASH	Has at least one internal dash
INITDASH	Starts with a dash
ENDDASH	Ends with a dash
SURROUNDPAREN	Surrounded by parentheses
HASPERCENT	Has at least one internal percent sign
PREFIX_i	Starts with the prefix of length $i \in \{2,3,4\}$
SUFFIX_i	Ends with the suffix of length $i \in \{2,3,4\}$
PUNCTUATION	Word is a punctuation
WORD	Word itself
WC	Word class
Extended features	
POS	Part-of-speech tag of the word
NPC	Word is part of a non-recursive noun phrase
WN_DISEASE	Word or phrase has hypernym Disease in WordNet
WN_DRUG	Word or phrase has hypernym Drug in WordNet
WN_PROTEIN	Word or phrase has hypernym Protein in WordNet
WN_DNA	Word or phrase has hypernym DNA in WordNet
WN_RNA	Word or phrase has hypernym RNA in WordNet
WN_CELL	Word or phrase has hypernym cell in WordNet
ISPLACE	Match word or phrase in region/country list

phrases as detected by the noun phrase chunker both for biological named entities and for named entities like drug and disease. For example, for named entities of type disease, the union of the set of hyponyms for WordNet synsets disease, mental illness, and genetic disease, which have only one or two word senses, forms the domain of this global semantic feature. The restriction in the number of word senses is influenced by the observation that the more common a word is, the more senses it has *(28)*. This way, the chance of selecting the wrong word sense is decreased by filtering out very common words. The word sense disambiguation scheme used for remaining words with two senses, is selecting the most common word sense.

3.4. Experimental Results and Analysis

For CRF training and labeling, as in ABNER, MALLET *(29)* is used. Performance is measured by standard IR measures, namely, precision, recall, and F_1 as defined by **Eqs. 12–14**. Here, counts are accumulated per named entity basis. For example, a named entity (NE) correctly identified and labeled increments the true positive count for that named entity type. All partial matches are counted as errors. The overall performance is measured by micro-averaged and macro-averaged precision, recall, and F_1 *(30)*. These overall measures are defined as

$$P^\mu = \frac{\sum_{i=1}^{C} TP_i}{\sum_{i=1}^{C} (TP_i + FP_i)} \quad R^\mu = \frac{\sum_{i=1}^{C} TP_i}{\sum_{i=1}^{C} (TP_i + FN_i)}$$

$$P^M = \frac{\sum_{i=1}^{C} P_i}{C} \qquad R^M = \frac{\sum_{i=1}^{C} R_i}{C}$$

$$F_1^\mu = \frac{2P^\mu R^\mu}{P^\mu + R^\mu} \qquad F_1^M = \frac{2P^M R^M}{P^M + R^M}$$

Here, C is the number of the classes/label types, being five and ten for datasets one and two, respectively. Macro averaging prefers performances of rare classes, while micro averaging prefers classes with over-average number of instances. In accordance with most information retrieval researchers *(30)*, micro-averaged measures are used as the main performance predictors.

For biological named entities, the first 5,000 sentences are used for training. For the second set the first 1,800 sentences of the 3,662-sentence corpus are used for training. To test if ABNER's state-of-the-art performance for biological named entities (F_1 around 70%) can be replicated in other domains, the first annotated set is labeled via ABNER. To compare ABNER and CRFNER, ABNER and CRFNER are both trained on the 5,000-sentence schizophrenia corpus. In **Table 5**, the results are summarized. The large drop in ABNER's performance shows

Table 5
NER performance results for biological named entities

NE	ABNER			ABNER Retrained			CRFNER+POS+NPC		
	P	R	F1	P	R	F1	P	R	F1
Micro-avg	40.94	58.95	48.32	62.15	43.33	51.06	68.21	46.61	55.38
Macro-avg	56.99	68.41	62.18	63	42.72	50.91	65.06	41.4	50.6
Cell_line	52.94	94.74	67.92	100	33.33	50	100	33.33	50
Dna	47.15	60.3	52.92	63.96	40.11	49.31	73.83	44.63	55.63
Rna	67.65	57.5	62.16	44.44	33.33	38.1	33.33	25	28.57
Protein	37.21	57.2	45.09	63.12	44.3	52.06	68.14	47.79	56.18
Cell_type	80	72.29	75.95	43.48	62.5	51.28	50	56.25	52.94

the importance of domain-specific corpora for training, which is especially apparent for the Protein NE class, constituting about 70% of all the biological NEs. When ABNER is retrained with schizophrenia corpus, its overall micro-averaged F_1 is slightly improved to 51.06%. Original ABNER is trained with a much larger dataset than ours, which may explain why its performance is better than both retrained ABNER and best CRFNER on cell_line, rna, and cell_type. Total number of cell_line, rna, and cell_type NEs make up less than 8% of the schizophrenia corpus. Overall the best-performing NER system for biological NEs is CRFNER with extended features POS tags and membership in a noun phrase. Part-of-speech tags act like a backoff mechanism for unseen words, allowing CRF to use POS occurrence statistics instead of missing word identity observation statistics, resulting in better generalization. Since most proteins and dna entity names are usually noun phrases, membership in a noun phrase will facilitate detection of these named entities in their entirety. WordNet features have slightly increased the overall performance (from baseline micro-averaged F_1 51.67–52.44%), but not as much as POS and NPC features. When combined with POS and NPC features, WordNet features decreased the performance below the baseline (micro-averaged F_1 51.07%).

For the second set of named entity types, CRFNER is trained with baseline features and different combinations of features from the extended set. The baseline performance and best-performing two combinations are summarized in **Table 6**. The performance improvement over mostly orthographic baseline features is modest at best. Use of location lexicon improved, especially recall, performance for the place NE. Since place NE size is small compared to the other NE types in this set, this was not reflected that much in the overall result. **Table 7** summarizes the effect of individual extended features and their important binary combinations. Semantic WordNet features slightly decrease overall performance and performance of drug and disease NEs compared to the baseline. By error analysis, this seems due to the overreliance on the WordNet feature which is a generic knowledge source sparse in biomedical areas and, especially for drug NEs, occasional incorrect word sense selection decision, e.g., using nouns with the addiction-causing substance sense of the drug.

4. Conclusion

In this chapter, two knowledge discovery support tools for neurodegenerative disease researchers are introduced. First, three new variants to Nigam and colleagues' *(1)* EM approach for document

Table 6
NER performance results for the second named-entity set

NE	Base			+LOC			+POS+LOC		
	P	R	F1	P	R	F1	P	R	F1
Micro-avg	88.26	76.9	82.19	88.4	77.3	82.48	87.98	77.35	82.33
Macro-avg	80.18	54.41	64.83	74.55	51.34	60.81	73.07	50.87	59.98
Percent	99.22	99.22	99.22	99.23	99.61	99.42	99.61	99.22	99.42
Dosage	88.24	53.57	66.67	94.12	57.14	71.11	88.24	53.57	66.67
Disease	90.74	86.67	88.66	90.5	86.39	88.4	90.44	86.62	88.49
Org	66.67	30	41.38	72.22	32.5	44.83	58.82	25	35.09
Time	67.7	61.05	64.21	67.97	61.05	64.33	66.92	61.05	63.85
Nationality	87.23	48.81	62.6	86.96	47.62	61.54	91.3	50	64.62
Age	50	23.08	31.58	50	23.08	31.58	55.56	25.64	35.09
Place	88.46	40.35	55.42	92.86	57.02	70.65	90.41	57.89	70.59
Assessment	78.57	34.59	48.03	80.3	33.33	47.11	78.57	34.59	48.03
Drug	84.95	66.76	74.76	85.92	67.04	75.32	83.87	65.92	73.82

Table 7
Performance effects of the extended features for the second named-entity set

NE	+WN F1	+POS F1	+NPC F1	+WN+NPC F1	+WN+POS F1	+WN+LOC F1
Micro-avg	81.84	82.15	81.87	81.13	82.07	81.6
Macro-avg	64.18	58.79	65.41	63.71	58.6	58.1
Percent	99.42	99.42	99.42	99.03	99.42	99.22
Dosage	66.67	66.67	71.11	63.64	66.67	73.91
Disease	88.28	88.63	88.34	87.7	88.72	88.05
Org	36.36	36.67	44.83	43.33	35.09	25
Time	64.18	63.35	61.85	63.33	63.02	64.71
Nationality	62.6	62.99	61.42	64.57	62.02	60.94
Age	31.03	31.03	32.14	27.59	31.03	32.14
Place	54.76	62.43	56.65	53.01	62.77	57.14
Assessment	48.21	48.48	48.95	48.93	49.35	48.65
Drug	73.46	74.52	74.8	71.78	73.23	72.4

classification from small number of labeled documents and larger set of unlabeled ones are introduced in search for a time-efficient search result refinement mechanism for BIRN Query Atlas meta-search engine. On average, all of the methods introduced have outperformed basic EM approach. MMEM(AIB) has shown better average F_1 performance, though not statistically significant, than the state-of-the-art transductive SVMs with a more than one order of magnitude improvement in execution time.

The search result refinement mechanism integrated with Query Atlas meta-search engine uses KNN for degenerate cases with no negative examples and switches to MMEM(AIB) if there are more than three positive examples. This conservative threshold on positive examples is chosen to ensure that Naive Bayes, the underlying algorithm for MMEM(AIB), always has enough examples for best performance.

Second, the NER tool introduced in this chapter, CRFNER, outperformed both original ABNER and its retrained version on the schizophrenia corpus, thanks to its extended set of features. For the second named entity set, CRFNER achieved 82.5% micro-averaged F_1 using baseline features and country/region lexicon.

As a complement to the direct relevance feedback-based approaches investigated in this chapter, user preferences can also be inferred from the search results the user has clicked. This partial ranking information can be used to rerank the search results so that the relevant search results appear before the irrelevant ones. Our further research on Query Atlas meta-search engine is focused on SVM-based search result relevance ranking function learning using click-through data *(31)*, to have a meta-search engine capable of exploiting both direct and indirect user feedback to provide a better search experience for the user.

On the NLP front, we are moving forward in building an annotated corpus for automatic classification of semantic labels in sentences of schizophrenia abstracts to extract semantic relations between entities of interest. Linking of semantic information between sentences is further investigated with the ultimate goal of a building a knowledge base for neurodegenerative disease literature capable of answering natural language questions.

5. Notes

1. Facts about MEDLINE are available at http://www.nlm.nih. gov/pubs/factsheets/medline.html

2. PubMed search engine is available at http://www.ncbi.nlm. nih.gov/entrez/query.fcgi

3. The Query Atlas meta-search engine is available at https:// loci.ucsd.edu/qametasearch/

Acknowledgment

This research was supported by 1 U24 RR021992 to the Function Biomedical Informatics Research Network (BIRN, http://www. nbirn.net) that is funded by the National Center for Research Resources (NCRR) at the National Institutes of Health (NIH). Special thanks to Sinem Özyurt MS Chem, MS ChE for annotating biological named-entities.

References

1. Nigam, K., McCallum, A. K., Thrun, S., and Mitchell, T. (2000) Text classification from labeled and unlabeled documents using EM. *Machine Learning*, **39**, 103–134.

2. Slonim, N., and Tishby, N. (1999) Agglomerative information bottleneck. In *Proceedings of NIPS-12*.

3. Davies, D. L., and Bouldin, D. W. (1979) A cluster separation measure. *IEEE Transactions on Pattern Analysis and Machine Intelligence*, **1**, 224–227.

4. Joachims, T. (1999) Transductive inference for text classification using support vector machines. In *Proceedings of the International Conference on Machine Learning (ICML)*.

5. Lafferty, J., McCallum, A., and Pereira, F. (2001) Conditional random fields: Probabilistic models for segmenting and labeling sequence data. In *Proceedings of ICML-2001*, 282–289.

6. Settles, B. (2004) Biomedical named entity recognition using conditional random fields and rich feature sets. In *Proceedings of COLING 2004 International Joint Workshop on Natural Language Processing in Biomedicine and its Applications (NLPBA)*.

7. Brown, G. G., Pieper, S., Martone, M., Aucoin, N., Joyner, A., Bischoff-Grethe, A., and Torvik, V. (2004) The query atlas: A brain referenced knowledge discovery tool. In *Annual Neuroscience mMeeting*.

8. Baeza-Yates, R., and Ribeiro-Neto, B. (1999) *Modern Information Retrieval*. ACM Press, New York.

9. van Rijsbergen, C. J. (1979) *Information Retrieval*. Butterworths, London.

10. Croft, W. B., Cronen-Townsend, S, and Lavrenko, V. (2001) Relevance feedback and personalization: A language modeling perspective. In *DELOS Workshop: Personalisation and Recommender Systems in Digital Libraries*.

11. van Rijsbergen, C. (1977) A theoretical basis for the use of co-occurrence data in information retrieval. *Journal of Documentation*, **33**(2), 106–119.

12. Friedman, J. H. (1997) On bias, variance, 0/1-loss, and the curse-of-dimensionality. *Data Mining and Knowledge Discovery*, **1** (1), 55–77.

13. Duda, R. O., Hart, P. E., and Stork, D. G. (2001) *Pattern Classification*. Second edition. Wiley, New York.

14. Dunn, J. C. (1974) Well separated clusters and optimal fuzzy partitions. *Journal of. Cybernetics*, **4**, 95–104.

15. Maulik, U., and Bandyopadhyay, S. (2002) Performance evaluation of some clustering algorithms and validity indices. *IEEE Transactions on Pattern Analysis and Machine Intelligence*, **24**(12), 1650–1654.

16. Gelatt, C., Kirkpatrick S., and Vecchi, M. (1983) Optimization by simulated annealing. *Science*, **220**, 671–680.

17. Ozyurt, I. B., and Brown, G. G. (2007) Search result refinement via machine learning from labeled-unlabeled data for meta-search. In *IEEE Symposium on Computational Intelligence and Data Mining CIDM 2007*, 186–193.

18. Levine, E., and Domany, E. (2001) Resampling method for unsupervised estimation of cluster validity. *Neural Computation*, **13**, 2573–2593.

19. Geman, S., and Geman, D. (1984) Stochastic relaxation, Gibbs distribution, and Bayesian restoration of images. *IEEE Transactions on Pattern Analysis and Machine Intelligence*, **6**, 721–741.

20. Peng, F., and McCallum, A. (2004) Accurate information extraction from research papers using conditional random fields. In *Proceedings of Human Language Technology Conference and North American Chapter of the Association for Computational Linguistics (HLT/NAACL-04)*.

21. Kindermann, R., and Snell, J. L. (1980) *Markov Random Fields and Their Applications*. American Mathematical Society, Providence.

22. Hammersley, J., and Clifford, P. (1971) Markov fields on finite graphs and lattices. Unpublished manuscript.

23. Sutton, C., and McCallum, A. (2006) An introduction to conditional random fields for relational learning. In *Introduction to Statistical Relational Learning* (Getoor, L., and Taskar, B., eds.). MIT Press, Cambridge.

24. Byrd, R. H., Nocedal, J., and Schnabel, R. B. (1994) Representations of quasi-Newton matrices and their use in limited memory methods. *Mathematical. Programming*, **63**(2), 129–156.

25. Charniak, E. (2000) A maximum-entropy-inspired parser. In *Proceedings of NAACL*, 132–139.

26. Ngai, G., and Florian, R. (2001) Transformation-based learning in the fast lane. In *Proceedings of North American ACL 2001*, 40–47.

27. Collins, M. (2002) Ranking algorithms for named-entity extraction: Boosting and the voted perceptron. In *Proceedings of Association for Computational Linguistics Conference*, 489–496.

28. Fellbaum, C. (ed.) (1998) *WordNet: An Electronic Lexical Database*. MIT Press, Cambridge.

29. McCallum, A. K. (2002) A machine learning for language toolkit. http://mallet.cs.umass.edu.

30. Sebastiani, F. (2002) Machine learning in automated text categorization. *ACM Computing. Surveys*, **34**(1), 1–47.

31. Joachims, T. (2002) Optimizing search engines using clickthrough data. *Proceedings of the Eighth ACM SIGKDD International Conference on Knowledge Discovery and Data Mining*, 133–142.

Chapter 10

Brain Model of Text Animation as a Data Mining Strategy

Tamara Astakhova and Vadim Astakhov

Summary

Imagination is the critical point in developing of realistic intelligence (AI) systems. One way to approach imagination would be simulation of its properties and operations. We developed two models "Brain Network Hierarchy of Languages," and "Semantical Holographic Calculus" and simulation system Script-Writer that emulate the process of imagination through an automatic animation of English texts. The purpose of this paper is to demonstrate the model and present "ScriptWriter" system http://nvo.sdsc.edu/NVO/JCSG/get_SRB_mime_file2.cgi//home/tamara.sdsc/test/demo.zip?F=/home/tamara.sdsc/test/demo.zip&M=application/x-gtar for simulation of the imagination.

Key words: Imagination, Text processing, Artificial intelligent, Animation

1. Introduction: Artificial Intelligence (AI) and Process of Imagination

Humans are exceptionally adept at integrating different perceptual signals to create new emergent structures, which results in new ways of thinking. Even in the absence of external stimulus, the brain can produce imaginative stimuli. Some of these imaginative stimuli are dreams and imaginative stories. The imaginative process is always at work in even the simplest construction of meaning, a concept that philosophers Gilles Fauconnier and Mark Turner call "two-sided blending" *(1)*. It is not hard to realize that the imagination is always at work in the subconscious. Consciousness usually views only a portion of what the mind is doing. Most specialists in various areas have impressive knowledge, but are also unaware of how they are thinking. And even though they are experts, they will not reach justifiable conclusions through introspection.

This led us to a conclusion that *imagination* is the crucial subconscious feature of the creative human mind and that it

Vadim Astakhov (ed.), *Biomedical Informatics,* Methods in Molecular Biology, vol. 569
DOI 10.1007/978-1-59745-524-4_10, © Humana Press, a part of Springer Science + Business Media, LLC 2009

might provide a basis for other mental functions. We should address this issue with respect to the problem of creating artificial intelligence. It is reasonable to state that a strong artificial intelligence system competitive with the creative human brain should exhibit a certain level of complexity. Rephrasing Searle (2) that artificial hearts do not have to be made of muscle tissue, whatever physical substance they are made of should have a causal complexity at least equal to actual heart tissue where the term "causal complexity" reflects the quantity of causal relations and their hierarchy. The same might be true for an "artificial brain" that might cause creativity and consciousness though it is made of something totally different than neurons, if the "artificial brain" structures share the level of causal complexity found in brains. It is not like building a "perpetuum mobile" as some people refer to when attempting to build AI. We do not observe a "perpetuum mobile" anywhere, though we can observe consciousness not just in humans. This observation can be made in other highly developed animals (3).

We think that the first step in making an artificial intelligence system conscious would be a mutual simulation of high-level human cognitive functions, such as memory and imagination with large-scale neural networks. Such a system will provide mapping between mental functions, combinations of the firing rate of the neurons, and the specific neuronal architecture; or even some biochemical features of the neuronal structures as suggested by Crick (4). The "Universal Grammar" optimal theory (5) and A Theory of Cerebral Cortex (6) demonstrate how discrete symbol structures can emerge from continuous dynamic systems such as neural networks. Those symbols can be represented as dynamic states over a set of distributed neurons where various symbols can be represented by various states in the same or different neural net. In our model of Hierarchy of Brain Network Languages, we proposed that a neural network's dynamics produce a hierarchy of communication "languages," starting from a simple signal level language, and advancing to levels where neural networks communicate by firing complex nested structures. Such communication is complex enough for syntactic structures to emerge from an optimization of neural network dynamics. We also provide a "Holographic Calculus" as a candidate for neural imprint computing that can lead to the emergence of semantic relations.

Before digging into our model, we would like to clarify that it is essential to distinguish between "primary consciousness," which means simple sensations, and perceptual experiences and higher-order consciousness, which include self-consciousness and language. We assume that the higher-order consciousness is built up out of processes that are already conscious, that have primary consciousness. In order to have primary consciousness, an AI should possess some mechanisms provided by the human brain.

Let us go through the list of features that should be implemented in an AI as a base.

Probably one of the most basic components is memory. The human brain is not just a passive process of storing memories, but is also an active process of recategorizing on the basis of previous categorizations. Adaptive Resonance Theory *(7)* is one of the candidates to provide a model of human memory. We modified ART and constructed a neural network that can store and categorize perceived images every time they are perceived rather than comparing them with stored templates. For example, if a child sees a cat, it acquires the cat category through the experience of seeing a cat and organizing its experience by way of the recurrent network [Shennon 8] first introduced by Shennon in 1948 and recently known in neuroscience by the term *reentrant maps* [Edelman 9]. Then the next time the child sees a cat, the child has a similar perceptual input. He or she recategorizes the input by enhancing the previously established categorization. The brain does this by changes in the population of synapses in the global mapping. It does not recall a stereotype but continually reinvents the category of cats. This concept of memory provides an alternative to the traditional idea of memory as a storehouse of knowledge and experience, and of remembering as a process of retrieval from the storehouse. It also explains the latest claims from recent psychological publications *(10)* why most of our memories of past events are constructed and have just a few correctly memorized elements. Based on recategorization, we can easily see how most of our memories can be constructed due to memory reinvention of each category from the most recent perceptual inputs.

Another critical component is the ability of the system to learn. The AI system has to prefer some things to others in order to learn. Learning is a matter of changes in behavior that are based on categorizations governed by positive and negative values.

The system also needs the ability to discriminate the self from the nonself. This is not yet self-consciousness, because it can be done without a discrete concept of the self. The system must be able to discriminate itself from the world. The apparatus for this distinction should provide the "body" system a set of spatial and temporal constraints, and should also register the system's internal states and discriminate them from those that take in signals from the external world, such as that feeling hunger is part of the "self," and the visual system, which enables us to see objects around us.

The AI needs a system for categorizing sequential events in time and for forming concepts. Not only should the AI be able to categorize cat and dogs, but it should also be able to categorize the sequence of events as a sequence. An example of this would be a cat followed by a dog. And it must be able to form prelinguistic concepts corresponding to these categories.

A special kind of memory is needed to mutually configure interactions among various systems. An example would be the experience of sunshine for warmth and the experience of snow for cold. The system should have categories corresponding to the sequences of events that cause warmth, or conversely, cold. And its memories are related to ongoing perceptual categorizations in real time.

We need a set of reentrant connections between the special memory system and the anatomical systems, which are dedicated to perceptual categorizations. It is the functioning of these reentrant connections that give us the sufficient conditions for the appearance of primary consciousness.

Using all of these features, we can define primary consciousness as an outcome of a recursively comparative memory, in which the previous self and nonself categorizations are continually related to present perceptual categorizations and their short-term succession, before such categorizations have become part of that memory.

On the other hand, higher-level mental functions should provide:

Conceptual integration is at the heart of imagination. It connects input spaces, projects selectively to a blended space, and develops emergent structures through composition, completion, and elaboration in the blend.

Emergent structures arise in the blends that are not copied directly from any input. They are generated in three ways: through projections composed from the inputs, through completion based on independently recruited frames and scenarios, and through elaboration.

Composition – blending can compose elements from the input spaces in order to provide relations that do not exist in the separate inputs.

Completion – we rarely realize the extent of background knowledge and structure we bring into a blend unconsciously.

Elaboration – we elaborate blends by treating them as simulations and running them imaginatively, according to the principles that have been established for the blend.

Another big area of human behavior involved in development of imagination is human internal "beliefs." The statement "person X believes Y" is equivalent to a series of conditional statements that assess how X would behave under certain circumstances. In that sense, beliefs are unobservable entities that cause observable traits in human behavior. These unobservable entities originated from facts of observation, from memory, from self-knowledge, and from experimentation. All of these blends in imagination lead to "a belief" through individual inference rules. Imagination is the result of merging and blending various concepts.

Following this list of high-level functions, we analyzed and implemented some aspects of imagination such as integration and identity in the software. We see integration as finding identities, and oppositions as parts of a more complicated process, which has elaborate conceptual properties that can be both structural and dynamic. It typically goes entirely unnoticed since it works so fast in the backstage of cognition. On the other hand, the identity is the recognition of identity and equivalence that can be mathematically represented as A = A. It is a spectacular product of complex, imaginative, and nonconsciousness work. Identity and nonidentity, equivalence and differences are apprehensible in consciousness and provide a natural beginning place for formal simulation approach. Identity and opposition are final products provided to consciousness after elaborate unconsciousness work and they are not a primitive starting point.

These operations are very complex and mostly unconscious for humans but at the same time play a basic role in the emergence of meaning and consciousness. From everyday experiences of meaning and human creativity, we can conclude that the meaning and basic consciousness operation lies in the complex emergent dynamics triggered in the imaginative mind. It seems reasonable to imply that consciousness and mind prompt for massive imaginative integration.

We believe that simulation of imagination is a first step for building a powerful AI system. To accomplish that step, a pluggable architecture called "ScriptWriter" was developed. That provides us with the ability to simulate imagination through the process of text animation.

2. Emergence of Holographic Network Processor and Holographic Calculus

A holographic network is organized as a graph-shaped hierarchy of nodes, where each node is a network itself and implements a common learning and memory function. **Figure 1** represents a process of development for the new holographic network.

First level of the holographic network is a set of sensory nodes. In a human, the optic nerve that carries information from the retina to the cortex consists of about one million fibers where each fiber carries information about light in a small part of visible space. The auditory nerve contains about 30,000 fibers, where each fiber carries information about sound in a small frequency range. The sensory node is like a fiber where each input noted measures a local and simple quantity.

Co-occurrence of signals on different sensory nodes leads to emergence of a so-called primary conceptual node that is

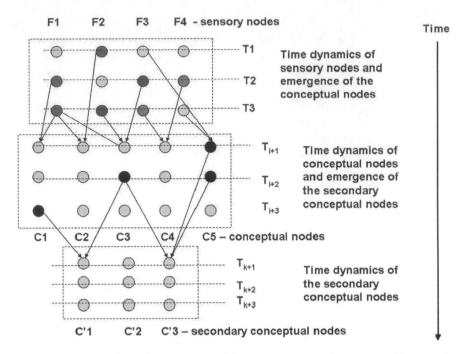

Fig. 1. Emergence of the holographic network represented in time as a process of development hierarchy of conceptual nodes through assembly of temporal signals in patterns called "concepts." Each concept it-self can participate in the process of development of new nodes.

implemented as a coherent firing of some subnet of neurons. same idea applied to the sequence of signals firing within a certain time window. The sequence of firing (T1, T2, and T3) or co-firing among sensory nodes will lead to emergence of a primary concept node as an assembly of those signals. In such a view, the primary concept is an internal representation of a spatial and temporal activity pattern of sensory nodes. The assembled pattern can be represented as a dynamic state of an underlying neural network.

Repeating of the same pattern with minor variations will increase the strength of underlying neural connections and enhance the strength of introduced concepts. By repeating the same pattern, the assembled concept node is getting *re-called* and updated. It stores the signature of the current pattern as well as previously observed. This way the concept keeps a history of the pattern evolution.

If variations of incoming new sensory patterns are sufficiently large, then the new primary concept is introduced. Those primary concepts can interact with each other and sensory concepts due to interactions among underplaying neural networks.

That network is a Bayesian network in which inferences represented by edges emerge as probability of nodes co-occurrence (be updated – accessed from sensory nodes) within a certain time window. We propose the formalism of those interactions through circular convolution and deconvolution which are eventually analogies of holography in optics. For any two random vectors X and Y, the circular convolution will produce another vector:

$$z = x @ y, \quad \text{where } z_j = \sum x_k * y_{j-k}$$

Convolution/deconvolution propagates as diffusion through network of concepts. Thus nodes interact (convolve/deconvolve) "holographically" with each other and can produce new imaginary nodes. same as for sensory nodes, the primary concepts will create a new layer of secondary concepts through the same mechanism of repeating sequences and co-occurrence within some time window. Those secondary, primary, and sensory nodes also can interact and re-currently lead to emergence of higher conceptual levels. Those interactions obviously create interconnections among nodes from various levels.

To not overextend a memory and keep only the significant experience, the intensity of each stored pattern signature exponentially decays. Each previously recorded pattern has a decay time dependent on the amount of secondary patterns emerged through co-occurrence with other concept during the time when the pattern was experienced.

Figure 2 illustrates how concept "C2" can stay longer in memory due to its participation in the emergence of several

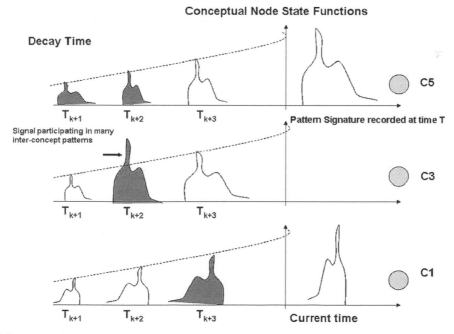

Fig. 2. Patterns recorded at different period of time represented by their signatures decaying in time. Figure represents reinforcement of the signature recorded at moment T(k + 2) for concept C3. Reinforcement performed due to C3 occurrence in emergence of new secondary conceptual nodes (**Fig. 1**) through assembly with C1 and C5 (**Fig. 1**).

secondary concepts "C′1 and C′2" through assembly with primary concepts "C5 and C1," respectively. **Figure 2** demonstrates *signatures* of concepts decaying in time. Those signatures are a numerical representation of dynamic states of underlying neural networks. If one concept participates in many assemblies which lead to the emergence of the new secondary nodes, then the concept gets reinforced each time the new secondary node is created or re-called.

If the node does not participate in the assembly of any other nodes, then it undergoes decay without reinforcement. The concept node keeps a signature as a vector in which coordinates represent recorded signals from sensory nodes multiplied by decay time exponent: $S1*\exp(-t/d)$. The decay time "d" will be represented as the amount of inter-node connections that will maximize the probability of the node to be re-enforced due-to various interactions.

When the intensity of the stored signature becomes less than some pre-defined threshold, then the signature vanishes from the holographic network. If all signatures recorded at different time for the concept vanish then the concept vanishes. This way the holographic network keeps itself adapted to current experience and eliminates old experience (signatures) and even old concepts. It also eliminates "conceptual noise" that is a bunch of new patterns that were stored and leads to the emergence of new primary concepts. Those concepts do not get any further support through repeating re-occurrences of their patterns. At the same as we demonstrate on **Fig. 2** the pattern participating in assembly of new concept nodes can be often reinforced by its assembly members and thus stay in "memory" even after a long period of non-recall.

We introduce conceptual node signature to represent dynamic states of the underlying network. **Figure 2** illustrates a signature as a one dimensional vector but it is actually a multi-dimensional complex (p-adic to keep order) vector. Each elementary signal from the sensory network is coded by set of neurons with different phases. **Figure 3** gives an imaginary analogy where each pixel of 2-D image can be represented as a concentric curve of another 2D image. We call such transformation a *delocalization*. That transforms all local features of the original image to distributed representation. It is very similar to the process of holography recording in optics. We call the new image a *holographic map*.

Each pixel on the new distributed representation of the original image can be seen as a neuron that keeps information about the original pixel and has some internal phase information that distinguishes its state from states of other neurons. Thus each sensory vector (matrix) has a delocalized holographic representation on the neural matrix that keeps local sensory information as a set of intensities and phases of distributed neurons.

**Original localized
representation**

**Delocalized Holographic
representation**

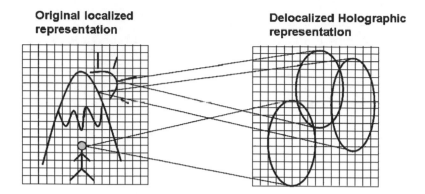

Fig. 3. Process delocalization through projection of each point of original image to a concentric curve of the new image.

Such representations let us realize circular convolution on neural networks and implement holographic calculus.

Inter-connection among neural networks provides a mechanism for concept reinforcement from other concepts. We introduce inter-connections among those neurons with scale free-distribution (**Fig. 4b**). Such architecture of the network will produce limited amount of "muster" nodes connected to all others. That topology will provide a light mechanism to re-call or reinforce the network through accessing only master neurons which will propagate the assessing signals to all others.

Figure 5 illustrates the process of reinforcement of some features of the old memories through convolution with new experience. Holographic representation of stored signatures that we called "holographic map" can be seen as some kind of "inner image." Categorization emerges through the reinforcements of different features of stored images. Such reinforcement is the result of assembly with other concepts or re-calls (update) from sensory nodes.

Convolution and deconvolution provide nice mechanisms for concepts and image operations. Any activity in a sensory or conceptual node will propagate through network of conceptions due to underlying network connections. Such propagation described by circular convolution and deconvolution is an analog of a holographic process (**Fig. 6**). That activity of certain nodes will be a result of convolutions that lead to emergence of a new set of nodes as a holographic reconstruction.

Propagation through those reconstructed nodes will lead to chain holographic reconstructions even further. This process will create waves of holographic reconstruction that are never ended and affected by external sensory node stimulation. Such holographic propagation leads to emergence of *prototypes* such as "tree" and "human."

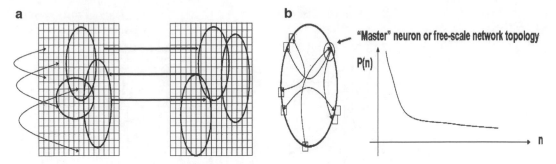

Fig. 4. (**a**) Interconnection among neurons forming holographic representation of sensor signals as well as connections among neurons from different holographic representations (maps). Re-entrant connections among holographic maps provide a mechanism for holographic convolution/de-convolution calculus. (**b**) Inter-connection among neurons involved in holographic representation of a local signal.

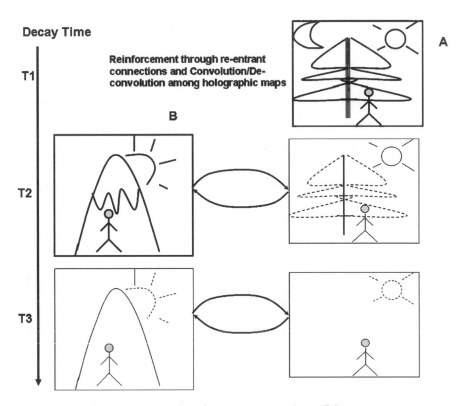

Fig. 5. Image/concept "A" stored in memory and reinforced by new experience "B."

Inner concepts continue the process of further convolution/ deconvolution among emerged concepts. Those concepts that will get higher reinformant through assembly with others or re-call from sensory nodes will survive and lead to emergence of prototypes.

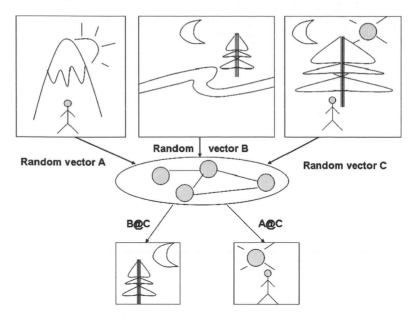

Fig. 6. Convolution/de-convolution among initial 2D-sensory signals into "internal concepts." Further convolution/de-convolution of concept signals into prototypes.

3. Ontological Model for Blending Process of Imagination

3.1. Concepts

We proposed "topological whole image computing" which based on whole image transformation and segmentation rather than starting from some elementary or primitive objects. Computing based on primitive concepts can be represented as symbolic computing but this "topological" approach deals with the nonlocal representation of a whole scene within the neural network of the AI-brain and integrated transformations over such representation. These transformations are neural networks, where primitive objects emerge as local invariants within a scene. We assume that, yet there is no supreme area or executive program binding the color, edge, form, and movement of an object into a coherent percept.

Objects represented as a whole in the juvenile AI-brain, as well as features such as color and shape, will diverge later over brain development.

A coherent perception in fact nevertheless emerges in various contexts, and explaining how this occurs constitutes the so-called binding problem.

The behavior of human infants conveys signs of strong synesthesia. So, we are suggesting that there is no "binding" in the juvenile brain, because it has not developed to the point where it can break perceptional fields into a set of modalities. For an adult brain the situation is different. A healthy and developed adult brain is well specialized for modalities like color, shape, and others.

The "topological" model takes binding as reentry mapping between distributed multimodel imprints from early childhood and features of a perceived field.

We propose to simulate mutual reentrant interactions among our holographic neuronal groups. For a time, various and linked neuronal groups in each map (neuronal group specialized for specific dynamics representing imprint, an object or some feature) to those of others to form a functional circuit. The neurons that yield such circuits fire more or less synchronously. (They provide a holographic representation of an object or an object feature.) But the next time, different neurons and neuronal groups may form a structurally different circuit, which nevertheless has the same output. And again, in the succeeding time period, a new circuit is formed using some of the same neurons, as well as completely new ones from a different group. These different circuits are degenerate, that is, they are different in structure yet they yield similar outputs to solve the binding problem. Such a multiple implementation can be realized in the holographic model.

As a result of reentry, the properties of synchrony and coherency allow more than one structure to give a similar output. As long as such degenerate operations occur in the correct sequence to link distributed populations of neuronal groups, there is no need for an executive program as there would be in a computer.

To construct an algorithm and a dynamic system, which emulates mental functions, fundamental theoretical units must be chosen. We propose the term "concept" for a sub-net of neurons exposing certain dynamical properties and at the same time is an internal representation of an object occupying space and time, an object with attributes specifying what an object is or does and what relations exist between objects. Example: concepts "woman," "walk," "beach" can lead to the conceptualization "A woman walked on the beach" that will lead to an animated image of a woman walking on the ocean beach. This conceptualization will imply many "beliefs." One possible belief here can be – "the woman wears something." Many of us intuitively "believe" that people usually wear something when they "walk" if the opposite is not mentioned. That "belief" will cause the imagination of many people to provide an image of the woman walking on the beach and wearing clothing, even if nothing was mentioned about clothing. It is a totally different case where we have the sentence: "A naked person is walking on the beach."

The following relevant rules can exist in the system:

IF X is like Y then X seeks Y.

IF Y disturbs X then X avoids Y.

But sometimes such rules can be in a conflict that leads to the emergence of new, blending structures.

3.2. Universal Structure of Objects in the Scene

A participant in the scene entity is assumed to be in one of three states (Active Actor, Passive Actor, and Action) with a binding pattern for every disclosed relation. Every entity's element (relations and attributes) has a descriptor for a keyword search, and a so-called semantic-type that can be used to map the element to its ontology. For example: Relation – behind (far behind) has a type-position/orientation. Another example: Attribute – red has type-color.

Further, an entity may disclose a set of functions that are internally treated as relations with the binding pattern (b, f), where b represents a set of bound arguments and the single f is the free output variable of the function. For example, "take the ball" can be treated as a human specialized function, which is used to raise the human actor hand in a set of specified scenes. Such functions will depend on sets of binding parameters "b" that they characterize, or the position of the ball and return "f"-position of the hand. Such a model lets us treat animation as Petri Net dynamics with computations where actors-nodes take different states in time.

3.3. Mental Space

We use the term "mental space" as small packets of concepts, which are constructed as we think and talk. Also, we have *conceptual integration* as a critical part of imagination. It connects input *mental spaces*, projects selectively to a blended imaginary space, and develops emergent structures through composition, competition, and elaboration in the blend.

For example, a set of sentences: "The blue ball was left on the beach. A woman walks on the beach," imply two input concept spaces "Woman walks on the beach" and "ball was left on the beach."

We perform *cross-space mapping* which connect counterparts in the input mental spaces and then construct *generic space* that maps onto each of the inputs and contains what the inputs have in common: beach, ocean, and horizon. The final *blending* does the projection of the ocean beach from the two input mental spaces to the same single beach in blended imagination space.

Such blending develops emergent imaginative structures that are not present in the inputs like "woman walk toward the ball" or "woman walk relatively close to the ball." It seems intuitively obvious that imagination can create an integrated scene with all the mentioned objects as a result of those two sentences.

3.4. Ontologies

Ontology is a term-graph whose nodes represent terms from a domain-specific vocabulary, and whose edges represent relations that also come from an interpreted vocabulary. The nodes and edges are typed according to a simple, commonly agreed upon set of types produced by testbed scientists. The most common interpretation is given by rules such as the transitivity of *is-a* or *has-a* relations, which can be used to implement inheritance and composition. However, there are also domain-specific rules for relationships such as *region-subpart* (rock-region→mountain-region) and *expressed-by* (emotion-state→face) that need special rules of inference. For example, if a rock-region participates in an imaginary scene (such as "he climbs the rock") and the human-emotion is expressed-by a face, then the rocks case is an emotion that will be expressed on the face.

In the current *ScriptWriter* framework, ontologies are represented as a set of relations comprised of a set of nodes and a set of edges, with appropriate attributes for each node and edge. Other operations, including graph functions such as path and descendant finding, and inference functions like finding transitive edges are implemented in Java.

We build ontology by extracting relations pair-wise between English words like "head – part of –body." We also assign a wait for each relation that reflects the probability to have two words in one sentence or in two concurrent sentences. That probability was extracted as a frequency of pair-wise occurrences of the two specified words. To perform a calculation, a test cohort of the fiction texts was collected.

3.4.1. Ontological dK-Series and dK-Graphs

Ontological graphs created the way described above are dependent on the cohort of the text and cannot pretend to be generic enough. Also, even for a small dictionary of English words this is extremely complex. Here, we need a way to approximate properties of a generic ontological graph, that can be built on a limited text cohort but that can capture topological properties of generic English text.

To capture such complexity of graph properties, we use the dK-graphs approach (*6*). This approach demonstrates that properties of almost any complex graph can be approximated by the random graph built by set of dK graphs:0K, 1K, 2K, and 3K where "K" is the notation for a node degree and d-for joint degree distribution that d node of degree "k" are connected.

Based on our text cohort, we estimate 0K-average node degrees as the average frequency of a word, 1K – node degree distribution as frequencies for the words in the text cohort, 2K and 3K – joint degree distributions were extracted as pair-wise and triple-wise frequencies of having two/three words in two/three subsequent sentences. Those values were assigned for each dK graph of our ontology and later used during operation of *ontological confabulation*.

3.4.2. Term-Object-Map

We have a specific source called the *term-object-map* that maintains a mapping between ontological terms and 3D-animation objects library, which was developed for the Maya animation environment. These objects are used by the system to build animation.

3.4.3. Mapping Relations

Currently in the animation industry, the burden of creating complex animated scenes over many actors is placed on the animation specialist, who works hard to capture the requirements of the script at hand. This leads to the pragmatic problem that the relationships between attributes disclosed by different objects and between object parameters are, quite often, not obvious. To account for this, the system has created additional *mapping relations*. Currently there are three kinds of mapping relations.

The *ontology-map* relation that maps data values from an object to a term of the ontology.

A joinable relation that links attributes from different objects if their attribute types, relations, and semantic types match.

The *value-map* relation which maps a fuzzy parameter value (speed fast) to the equivalent attribute value disclosed by the animation software.

4. Universal Text Filtering and dK-Ontology Confabulation

4.1. Text Filtering

ScriptWriter uses simple English text as an input and generates output animation. First, it performs text processing to extract semantic relations among words in the sentences. Mental space is created for each sentence. As an example, we consider the simple text of three sentences: "*A woman walks on the beach. The blue ball was left on the beach. A woman takes this ball.*"

Figure 7 represents details of the first mental space are built from the sentence "Woman walks on the beach." The sentence was processed and its *universal structure* was extracted: "Active actor (woman)–action (walk)-passive actor (beach)." Instances of the universal structure (woman, walk, and beach) were anchored (colored by yellow) to an ontological graph that represents relations among concepts. We perform graph expansion operations to integrate all relevant objects required for the mental space such as ocean, sky, and the woman's clothing, which are not mentioned in the sentence (colored by red).

Each concept such as "beach," "woman," and "ball" represents a subgraph that connects all concepts relevant to the specified term.

The same operations were performed for the second and third sentences.

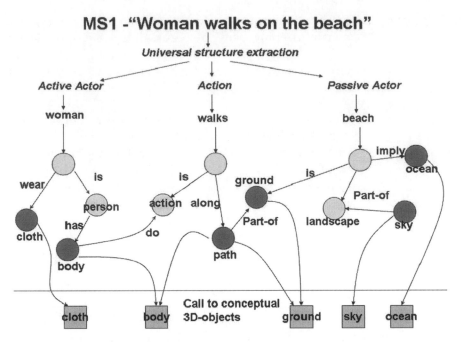

Fig. 7. Part of ontological graph that represents mental space for sentence "Woman walks on the beach."

4.2. Graph Calculus and Generic Space

Generic space was constructed by the mapping of objects such as "woman," "beach," and "ball," which co-occurred in different *mental space*.

Starting from those concepts we perform node expansion as a procedure of finding neighbor concepts connected to generic concepts; for example, the "color" of the ball and "body" of the woman. We perform node expansion by using various relations (represented by edges on the ontological graph) such as analogy/disanalogy, cause-effect, representation, identity, part-whole, uniqueness, similarity, and various properties. All these classes of relations are represented by various grammar constraints in the English texts. Following the ideas of "Universal Grammar" and "Distributed Reduced Representation" proposed in papers of Paul Smolensky *(8)*, we define any semantic space as a convolution of some vectors-concepts providing the space reduced representations. Rather than using Smolensky's Tensor *(7)* that has the variable length, we decided to use the holographic reduced representation techniques *(8)*.

Each node and edge were assigned to a random vector from 512-dimensional space, then any combinations of connected nodes and edges were defined as a result of circular convolution on the vectors *(8)*: $z = x \oslash y$, where $z_j = \sum x_k {}^* y_{j-k}$.

And indexes represent coordinates in 512-dimensional space. Such representation will provide coding schema to any complex scene.

Example: subgraph "woman-wear-clothing" was represented as z = x @ y = x @(m @ n); where "m" – represents vector "woman," "n" – represents "wear," "clothing" and convolution of "m" and "n" gave us "woman-wear-" open-end subgraph that has one node and one edge.

4.3. Renormalization as Dimension Reduction

Due to the high complexity of the ontological graph we performed a "graph compression" that resulted in elimination of some redundant links. We decided to perform compressions just over the vital relations such as Time, Space, Identity, Role, Cause-Effect, Change, Intentionality, Representations, and Attributes. Mathematically, that operation was implemented as circular correlation: $y = x \# z$, where $y_j = \sum x_k {}^* z_{k+j}$. Taking the previous example, that operation is equivalent to

"woman-wear-"# "woman-wear-clothing" = "clothing" and will return the node "clothing."

Holographic reduced representation lets us quickly compute "generic space."

The final ontological subgraphs blending and generation of resulting imaged space were performed as a confabulation on the *generic space.*

4.4. dK-Ontological Confabulation

We extended the operation of Confabulation previously proposed in paper *(7)* and developed an extended version of this operation for ontological graphs. This operation extracts concepts not mentioned in the text message. Consider our example: "A blue ball was on the beach. A woman walks on the beach. She takes the ball and kicks it." It seems clear that our imagination should build the picture of the woman that binds her body to take the ball by her hands, even though nothing in the text mentioned that biomechanical process. To do that, *Imaginizer* will use *ontological confabulation* for extracting knowledge associated with provided concepts. Confabulation was defined *(9)* as a maximization of probability to start from nodes a, b, and c and get node d: p(abc|d) ~p(a|d)* p(b|d)* p(c|d).

We start from any input node A, and then randomly walk and calculate its probability to get to node B. That probability obviously depends on the order degree for node B. The more nodes connected to B through some path, the higher the probability it is to get there. The initial algorithm proposed in *(9)* uses only the weight of the edges that were calculated from pairwise frequencies of two nodes in some text.

Here we propose a new algorithm to calculate the probability of transitions by using node degrees and joint probabilities of dK – series (0K, 1K, 2K, 3K) that was extracted from the cohort of the text during building of the ontological graph. Due to analysis **(6)** that any properties of the complex graph can be reconstructed by a random graph with identical statistical properties for 0–3K subgraphs, we suggest calculating the probability of a transition from A to B through some intermediate nodes as a sum over degree distributions for all intermediate 0K, 1K, 2K, and 3K subgraphs between A and B.

If we start from several input nodes, then the total probability to get to node B is the sum over all probabilities calculated for each input node. The B node with highest probability will be taken as a part of new blended space. The next less probable node was taken as a part of the imagined space if its probability was higher than some threshold. That threshold was estimated heuristically.

Each scene is represented as a vector as well as graph of relevant concepts. Dynamics of underlying neural groups can implement various operations on graphs. Those operations can be illustrated using holographic analogy from optics. **Figure 8** illustrates deconvolution of the blended space with two vectors (Subscene 1 and Subscene 2) which represent various subscenes. The result vectors are called "holography" due to similarity between mathematical formalism for our graph operators and optical holography. The result of deconvolution is holographic reduced representation for some portion of our blended graph. Those portions obviously depend on the deconvolution vectors. That is illustrated in **Fig. 9** where each resulting vector encodes a graph of concepts related with each. Encoded graphs are internal representation of visual scenes in the "ScriptWriter brain."

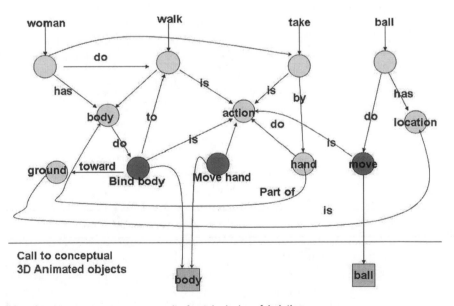

Fig. 8. Final imaginary space emerges as a result of ontological confabulation.

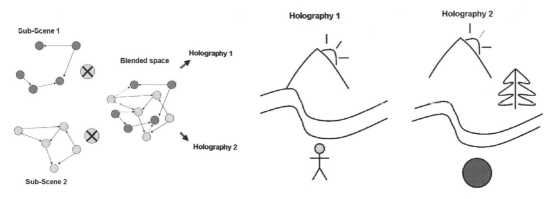

Fig. 9. Part of ontological graph that represents mental space for sentence "Woman takes the ball."

5. "ScriptWriter" and Animation Process

ScriptWriter is an attempt to simulate visual imagination through the creation of 3-D animation of English text. It is done in such a way as to approximate a human reading a story and imagining it from a first-person perspective.

The first limited version of our software called *ScriptWriter* provides us with the ability to "imagine" short stories with primitive objects such as humans, actors, and several landscapes. ScriptWriter performs text processing, semantic extraction, and animation planning that utilize approaches tested in various areas of Interactive Virtual Environment development *(2–5)*. Those were designed specifically for the development of believable agents – characters which express rich personality, and which, in our case, play roles in an imaged animated world. ScriptWriter platform provides a set of libraries and APIs. We propose a Concept Mapping tool which integrates ontology with mapped objects stored in a relational database. The ontology is described using OWL (Ontology Web Language) which is built on top of RDF (Resource Description Framework), which can be edited using a tool such as Protégé. An OWL ontology is constructed with a hierarchical vocabulary of terms to describe concepts in blended space. Each concept in the ontology maps to a Data Object, which is a set of fields and values stored in a database. There is also the capability to retrieve related data fields via sequences of foreign key/primary keys and map those field values. The full set of concept–data object mappings is saved to a project which can be loaded into a database. It is then possible to perform a join query using concept terms to retrieve all relevant data stored in the database as a result set. This method allows ontological concepts to be used as a generalized query vocabulary for database information retrieval.

It provides a connection to sensory-motor system of agents – "Actors" and supports multi-agent coordination. ScriptWriter scenario was organized as a collection of the behavioral actions of the actors or simply "Actions." During the imagination process some of the images and their actions encapsulate some other actions in the same kind of nested way that will produce sequential behavior. An example of sequential behavior is: "A woman walks on the beach. There was a blue ball on the beach. She kicks the ball." We easily can imagine a scene where a woman walks to the ball left on the beach by someone and how she is kicking the ball. The text can be animated now by ScriptWriter with minimal human intervention.

6. Notes

The demonstration **Fig. 10** presents the "ScriptWriter" application performing animation for short texts of two to four sentences. This will include a demonstration of text processing and building

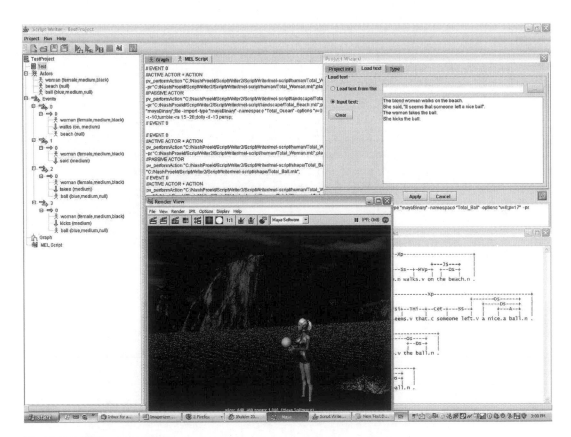

Fig. 10. ScriptWriter screen shot.

semantic relations as well as a generation of a scenario for the animation. This scenario will be used to automatically generate a script for building animation in the "Maya" animation environment.

This project raises a number of interesting AI research issues, including imagination process management for coordinating visual object interactivity, natural language understanding, and autonomous agents (objects, landscapes, and their interactions) in the context of a story. These issues were partially answered in the current version and will be refined in the next generations of the software.

We also suggest simulation of the new network processor architectures based on proposed holographic calculus.

Acknowledgment

We thank Edward Ross and David Little from University of California San Diego for discussion and comments.

References

1. Fauconnier, G. and Turner Mark Book: The way we think. 2003CHI Conference Publications Format. Available at http://www.acm.org/sigchi/chipubform/.

2. Bates, J. 1992. Virtual Reality, Art, and Entertainment. *Presence: The Journal of Teleoperators and Virtual Environments* 1(1): 133–138.

3. Bates, J., Loyall, A. B. and Reilly, W. S. 1992. Integrating Reactivity, Goals, and Emotion in a Broad Agent. In Proceedings of the Fourteenth Annual Conference of the Cognitive Science Society. Bloomington, IN, July.

4. Mateas, M. and Stern, A. 2000. Towards Integrating Plot and Character for Interactive Drama. In *Working Notes of the Social Intelligent Agents: The Human in the Loop Symposium*. AAAI Fall Symposium Series. Menlo Park, CA: AAAI Press.

5. Sengers, P. 1998. Do the Thing Right: An Architecture for Action-Expression. In *Proceedings of the Second International Conference on Autonomous Agents*. pp. 24–31.

6. Mahadevan, P., Krioukov, D., Fall, K. and Vahdat, A. A Basis for Systematic Analysis of Network Topologies. http://www.krioukov.net/~dima/pub/index.html.

7. Prince, A. and Smolensky, P. 1997. Optimality: From Neural Networks to Universal Grammar. Science 275: 1604–1610.

8. Tony, A. 1997. Plate Holographic Reduced Representation. CSLI Lecture Notes No. 150.

9. Rumelhart, D. E., McClelland, J. L. and the PDP Research Group. *Parallel Distributed Processing*. Vols. 1 and 2. Cambridge, MA: MIT Press.

10. Hecht-Nielsen, R. A Theory of Cerebral Cortex. UCSD Institute for Neural Computation Technical Report #0404.

Chapter 11

Processes Parallel Execution Using Grid Wizard Enterprise

Marco Ruiz

Summary

The field of high-performance computing (HPC) has provided a wide array of strategies for supplying additional computing power to the goal of reducing the total "clock time" required to complete various computational processes. These strategies range from the development of higher-performance hardware to the assembly of large networks of commodity computers, with each strategy designed to address a particular aspect and/or manifestation of a given computational problem. GWE (Grid Wizard Enterprise) in that regard, is an HPC distributed enterprise system, aimed at providing a solution to the particular problem of running inter-independent computational processes faster by parallelizing their execution across a virtual grid of computational resources with a minimum of user intervention.

Key words: Cluster computing, Grid computing, gwe, High performance computing, Parallel computing, Resource manager, Job scheduler, Globus, Java, Open source

1. Introduction

Engineering and technology has always faced the relentless demand from society to come up with better, cheaper, and more efficient solutions for its many problems. This is especially true for HPC technology; which is in charge of providing solutions for the problem of supplying high computation power.

The domain problem HPC addresses is the one that arises when a user requires to run "some" process over "some" data; and the computation power available requires an unacceptable amount of time to fulfill such requirement; due to a combination of a high level of complexity of the processing algorithms and a high amount of data to process. HPC technology, in its relatively short lifespan, has come up with a great variety of strategies to overcome this problem; strategies which range from

Vadim Astakhov (ed.), *Biomedical Informatics,* Methods in Molecular Biology, vol. 569
DOI 10.1007/978-1-59745-524-4_11, © Humana Press, a part of Springer Science+Business Media, LLC 2009

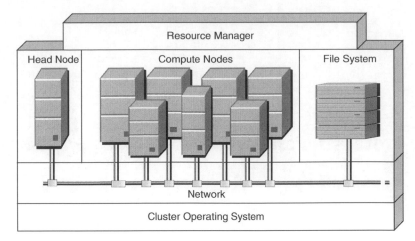

Fig. 1. Cluster anatomy.

more efficient processors and memory architectures to cluster, cloud, and grid computing. As with many technology realms, in the HPC many of these technologies leverage on others.

Cluster computing is an HPC technology; which aims at encapsulating a number of computers in order to create the equivalent of a single virtual supercomputer (from the end user's point of view). In order to do so (to "glue" these computers) many standard technologies are needed: a higher speed network to interconnect the computers, a higher speed shareable data infrastructure (database and/or shared file system), a specialized operating system, and a designated specialized computer which will serve as the coordinator of the cluster, known as the cluster head node (*see* **Fig. 1**).

On top of this cluster infrastructure many process parallelization technologies are evolving but one has gathered enough standardization for users to start counting and building on top of it: the "resource manager." This system (typically running in the cluster head node), at its least, provides functionality to queue and execute independent processes in the cluster. Today, there are many implementations of such type of system (Condor, SGE, LSF, PBS Torque); all of them differing widely in behavior and feature set; but remaining with the constant of queuing and executing processes in a cluster; but still leaving much to the end user to deal with before having a straightforward platform for parallelizing its processes.

2. Processes Parallelization Problem

With the cluster plus resource managers infrastructure one would think that the end user has an easy-to-use platform to parallelize the execution of its processes; but unless its processes are extremely straight forward and trivial, the end user faces a daunting challenge.

Most real processes parallelization problems require resolution of issues, which, even in the presence of a powerful resource manager, have to be resolved by the end user. Among the most critical ones we have:

1. Upload the data to be processed to the cluster (localization).

2. Submit all processes to compute nodes (queue jobs in resource managers).

3. Monitor processes execution progress (real time and querying on demand).

4. Send/receive custom alert notifications (certain interesting conditions reached such as a percentage of processes completed execution).

5. Failover and recovery from cluster and environment related problems.

6. Failover and recovery from processes related problems.

7. Gathering and compilation of processing results.

8. Uploading result data to the storage resource of your choice.

9. Cleaning up the original and result data from the cluster data storage resource.

But that is not all; clusters are, typically, shared multi-user organizational resources and cannot be monopolized; this means that every single one of the previous issues became far more complicated to overcome; and new ones (such as coordination of user data workspaces, security for data and processes, prioritization, multi-user administration, etc.) expand the list. This solution is far from straightforward and requires too much from the end user to consider it practical; which is why process parallelization has not become mainstream yet.

Two different tendencies are evolving in the industry, as a way to bridge this gap, with (in the author's opinion) negative consequences: users gathering a considerable level of technical knowledge, which is taking away time and effort from their actual domain problems; and the creation (by technically savvy users and/or IT departments) of highly customized scripts and applications tailored to specific parallelization problems, which makes it harder to move toward a standardization in the industry for this problem: instead of evolving toward an inorganic cohesive solution, which addresses all related issues to the problem, this approach maintains the organic, disconnected evolution of a solution.

3. Grid Wizard Enterprise (GWE)

GWE is an HPC distributed enterprise system, which leverages on cluster computing and resource managers in order to become a true practical solution for end users to easily and effectively parallelize the execution of their processes in cluster environments.

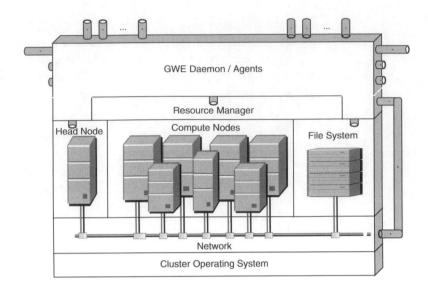

Fig. 2. GWE enabled cluster anatomy.

This product tries to achieve such a goal by addressing all issues related to process parallelization under a cohesive modular infrastructure that allows custom extensibility as well as providing a default implementation for most of these issues.

Users can get up and running in minutes, submitting their processes invocations with generalized tools that can easily connect to GWE enabled clusters (*see* **Fig. 2**). Such clusters (GWE daemons) will queue, prioritize, and resolve compute resources to execute these processes and finally execute them; keeping track (in an internal embedded database) at all times of all phases each process will go through providing the end user with a great wealth of information and control over their processes.

3.1. Requirements

As part of the philosophy of maximizing power and minimizing requirements, GWE has been implemented in a way that it only imposes two requirements on the clusters in which it is intended to run:

- The cluster and workstations must have Java 1.5 or greater available.

- The cluster must be SSH enabled.

3.2. Distributed System

As a distributed system, GWE requires many applications to be set up and running in order to work. The following are the subsystems; that compose the GWE system:

- *GWE daemon*. System running on the head node of a cluster. A cluster that has this process running on its head node will be referred to as a "GWE enabled cluster" and it is in charge of coordinating and relaying decisions across the GWE grids it belongs to (*see* **Fig. 3**).

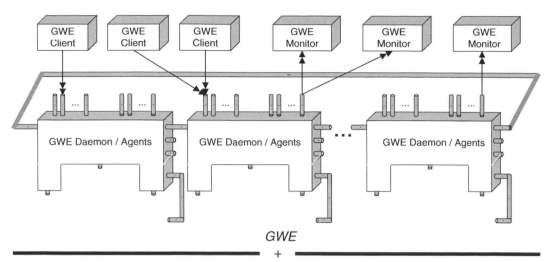

GWE
+
Clusters, Resource Managers and Network Enabled Devices

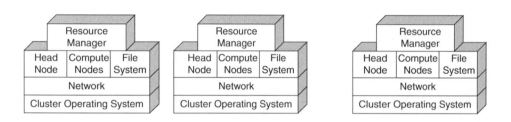

= GWE Grid

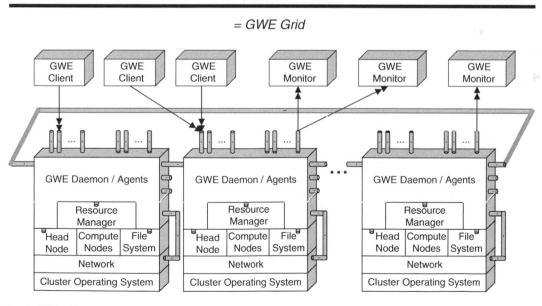

Fig. 3. GWE grid anatomy.

- *GWE agent.* Small footprint system running on each "allo-cated" compute resource of a GWE enabled cluster in order to serve as a proxy controller of such resource for the respective GWE daemon.

- *GWE client/monitor.* Any application built using the "GWE Client API" in order to allow end users to interact with GWE daemons.

Communications between all these subsystems are carried on using a secured RPC backbone built using RMI over SSH tunnels.

3.3. Features

Besides the main functional, end user, features (which will be described in the "User Types") section, GWE comes with a wealth of other features, which can be categorized according to value added, scalability and extensibility, and overall product robustness. The following is a list of these features:

3.3.1. Value Added

These are features, which are not necessarily required for process parallelization; but provide aid to common concerns of process parallelization:

- Simple XML-based configuration.

- Powerful language (P2EL) to describe process parallelization invocations.

- Administrator utilities to remotely deploy, configure, and launch GWE daemons.

- Resource manager auto discovery.

- Transparent download of input files, as long as the user provides their remote location and the authentication information needed to access it.

- Transparent upload of output files, as long as the user provides their remote location and the authentication information needed to write those files.

- Application management framework to allow users to auto-deploy their applications in the clusters, so they can be invoked by their orders.

- System events generation and logging.

- Alert definition/notification module to define and get alerts when custom interesting conditions arise.

- Embedded, proprietary GWE resource manager for clusters.

3.3.2. Scalability

These are features to be exploited by developers to create their own custom implementations of different GWE components:

- Rich client API to create custom GWE clients like the ones described in **Subheading 3.5**.

- Abstract job descriptors to create custom programmatically describable collection of parallelizable processes. P2EL is built on top of this framework.

- Abstract job result parsers to create custom parsers for process results.

- Parametrizable orders to provide custom data for the processes and behavior that should be invoked at the different phases a process goes through according to the order's callback signature.

- Driver management framework to accept new thin drivers to manage different grid related types of resources. These resource types are:

File Systems. Out-of-the-box support: Local, HTTP, and SFTP file systems.

Network Protocols. Out-of-the-box support: Local and SSH.

Resource Managers. Out-of-the-box support: Condor, SGE, PBS.

3.3.3. Robustness

These are internal, low-level features, which are meant to give GWE the robustness level expected from a true industrial-strength enterprise system:

- High-performance, secured RPC network backbone built using Java RMI over SSH tunnels used to transmit all GWE inter-subsystems communications.

- Scalable network design, which allows GWE daemons to daisy chain.

- Secured user sandbox where users information and processes parallelization do not interfere among each other in such a multi-user environment.

- Internal data warehousing through embedded database structurally logging all progress and events occurring in the system.

- Highly multithreaded to achieve maximum performance. Most possible wait cycles have been isolated using thread this and great care has been taken to guarantee the system is thread safe (no deadlocks or other anomalies appear).

- Automatic failover and recovery from cluster-and environment-related problems.

- Automatic failover and recovery from processes-related problems.

3.4. Enterprise System: User Types

As an enterprise system, it works with the concept of having multiple types of users, which for ease of usage have been limited right now to two: administrators and end users. This scheme is analogous to Unix-based systems with the root users and regular users.

3.4.1. Administrators

Administrator users are those who install (*see* **Fig. 4**) and manage a particular GWE daemon, using a valid SSH account,

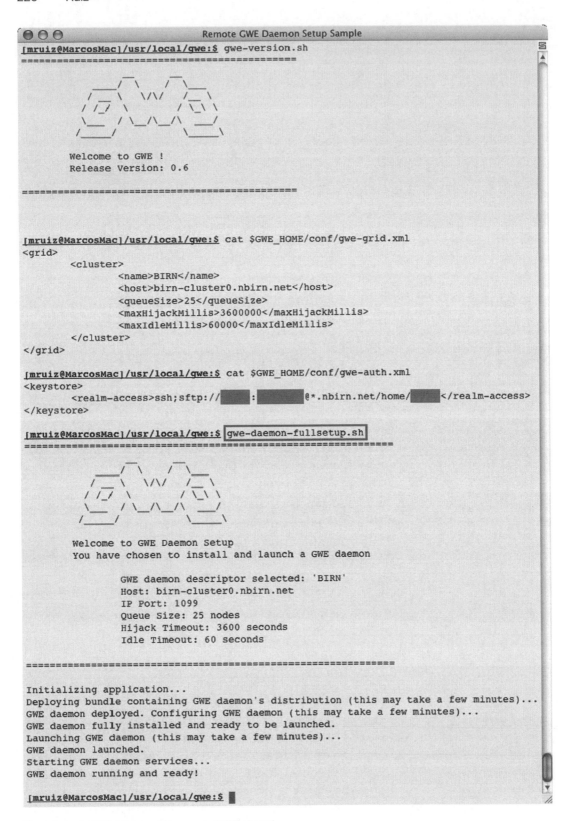

Fig. 4. Remote GWE daemon setup sample (BIRN cluster).

on the cluster head node. These users are responsible to carry on the following tasks over their GWE daemons:

- Deploy the GWE daemon into the head node of a cluster.

- Configure the GWE daemon with behavioral parameters such as queue size (maximum compute nodes allocated at any given time), maximum "hijack" time (maximum time a compute node can remain allocated), and maximum "idle" time (maximum time a compute node can remain allocated with nothing to do).

- Launch and shutdown when required.

- Provide an SSH account to communicate with cluster compute resources. By default it will use the SSH account used to install it.

3.4.2. End Users

End users are those who are able to connect to a particular GWE daemon being allowed to carry on the following tasks:

- Describe their own virtual grids as a collection of clusters the user has access to and has a GWE daemon running in them.

- Provide authentication information to GWE clients to access GWE daemons on the user's behalf.

- Provide authentication information to GWE daemons to access networked "grid resources" (such as remote file systems for reading/writing files) on the user behalf.

- Submit orders (collection of parallelizable processes) to a described grid for execution.

- Manage the execution of their orders (pause, resume, abort).

- Monitor the execution of their orders (real-time or on demand).

In its most generic form, GWE provides console user interfaces to interact with a GWE daemon and carry on these tasks (*see* **Fig. 5**).

P2EL (Processes Parallel Execution Language)

The generic process parallel execution requests are built using a GWE proprietary language (baptized as P2EL) to describe the order (collection of processes) to be executed in parallel. This language is a mix of pseudo bash and pseudo VLT (Velocity Template Language) – which are standard languages. This language adds its own set of powerful semantics to define:

- *Iteration variables.* Substitution variables and their respective range value, that will be used to generate a value-space set, which when applied as variable to value substitution to the template portion of a P2EL statement, generates the actual collection of processes invocations; which the statement represents. The semantics for iteration variables also include stepping information.

- *Numeric variable values pattern formatting.* Formatting and precision patterns to apply when doing value substitution of numeric values.

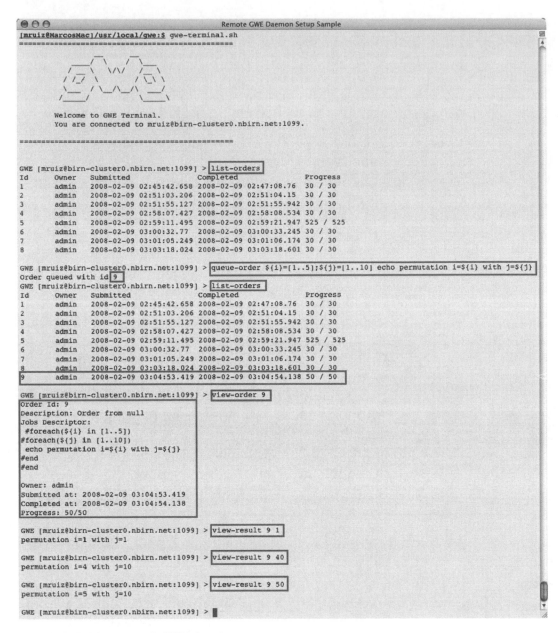

Fig. 5. GWE client usage sample (BIRN cluster).

- *Pre-interpreted wildcards.* Contextual value ranges for variables, resolved at runtime based on the context in which it will be running.

- *Process invocation template.* Template that will be used to create all the processes invocations the statement represents.

- *Substitution expressions.* These are "bash like" variable expressions embedded in the template; to be replaced by the corresponding value-space.

- *Remote file locations.* Instructions on how to locate remote files required to execute a process and locations of desired destinations for output files. By default the remote files will be used locally as files which names will match those of the last part of the URI. P2EL provides a means to override this default behavior by supporting semantics to provide a new name for automatic downloaded files.

3.5. Integration Case Studies

GWE comes with tightly integrated tools and application for specific third-party tools to provide them with grid computing enabling features.

3.5.1. Use Case: 3D Slicer

NA-MIC's "3D Slicer" is a rich client application that provides imaging processing features to dynamically pluggable modules. Such modules are regular applications, which must conform to a straightforward, proprietary specification. This specification requires each module to work normally unless invoked with predefined 3D Slicer reactors.

To integrate GWE with "3D Slicer" a small utility was developed to create a proxy module for each module discovered in a "3D Slicer" installation. Each of these modules is basically a Slicer module and a GWE client, proxying user's invocations to the actual modules and accommodating their responses according to GWE logic in order to execute batches of imaging processing in the cluster environment of a user's choice.

3.5.2. Use Case: LDDMM through BIRN Portal

BIRN's Portal is a web portal application for the biomedical community mainly focused on providing a grid data repository (SRB), and moving toward providing grid computing for selected applications. The first application selected is LDDMM, an image processing application, which in the BIRN portal should work off files stored in the SRB file system.

To integrate GWE with "LDDMM" an SRB file system driver was developed along with a custom portlet GWE client for the portal interface and custom order hooks to provide specific parameters to LDDMM type of processes.

3.6. Architecture and Design

GWE has been architected on top of the "Spring" application framework as a set of independent modules and frameworks declaratively glued using Spring's dependency injection. Such modules have been designed with a robust and scalable infrastructure and with multiple levels of abstraction to provide a balance between power and ease of usage across the board: from value added features, tools, and utilities for end users to application programming interfaces (API) for third-party tool integrators.

Most components have been built on top of abstract signatures (interfaces), allowing their implementations to be easily replaced and integrated into the system. For most components this will be unnecessary; but the driver manager framework is a perfect example of how a developer can leverage GWE by providing new drivers for file systems, network protocols, and/or cluster resource managers.

The following sections will address high-level details of the architecture and design of GWE subsystems in terms of subcomponents (modules and frameworks). Each of these components has been designed with a high level of self-containment, which allows them to be highly reusable. That is why across the subsystems many components can be repeated.

The architecture diagrams included in the subsections have a bolded border; which is used to indicate the scope of the subsystem and its components in terms of interfacing just with components inside the subsystem or if it also can/must interface with systems out of the scope of the subsystem.

3.6.1. Common Components

These are components that can be found in all GWE subsystems.

- *Secured RPC backbone.* A robust RPC module composed by a set of programmatic RMI remote APIs, a custom RMI layer tweaked to run over SSH tunnels, and SSH tunnels. A user of this layer would simply request an RMI server given its network address to initiate RPC type of communication with the remote system.

- *Persistent model.* The bare data-warehousing model of the system. Includes metadata to support persistence features when exposed to a DAO layer.

- *Driver manager framework.* A generalized framework that allows drivers to be used to request for resource handlers. Specific frameworks (of this generalized framework) have been defined to handle file system drivers, network protocol drivers, and resource manager drivers.

3.6.2. GWE Daemon Components (Fig. 6)

- *RMI servers.* SSH tunneled listening servers to provide services to other GWE subsystems.

- *Resource manager drivers.* This specific driver manager framework deserves a special mention because of its particular useful design automatic discovery feature.

- *Embedded database.* Derby, a Java relational database, has been integrated into GWE daemon as its persistent data-warehouse module.

- *DAO + ORM layer.* Data warehouse layer. The DAO layer has been created using generics for high reusability and the ORM layer has been implemented using Hibernate with Spring's declarative transaction management.

- *External processes.* These are all services a GWE daemon provides to its clients, agents, and internal processes (queue orders for

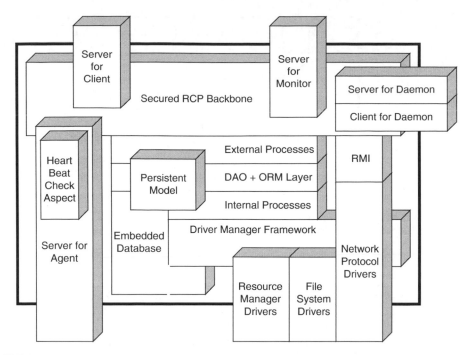

Fig. 6. GWE daemon architecture.

future execution, report status information as required, request execution of processes to agents, etc.).

- *Internal processes.* These are all internal services a GWE daemon provides to external processes (allocation of compute nodes, recycling allocations, downloading and uploading files, cleaning up workspaces, logging events, etc.).

3.6.3. GWE Agent Components (Fig. 7)

- *Abstract daemon request processor.* A module that runs a daemon request on behalf of the GWE daemon in the compute node it is running. Particular implementations of this processor include system routines for self-diagnostics, agent orderly shutdown, and the most common process parallelization.

- *Heart beat aspect.* Fault tolerance and recovery type of component. An aspect which periodically emits a health status update message to its GWE daemon.

3.6.4. GWE Client Components (Fig. 8)

- *Client for client.* An RMI client connected to an instance of the RMI "Client API" server of a particular GWE daemon to interact with it.

- *Client for monitor.* An RMI client connected to an instance of the RMI "Monitor API" server of a particular GWE daemon to retrieve events from it. This connection has an internal

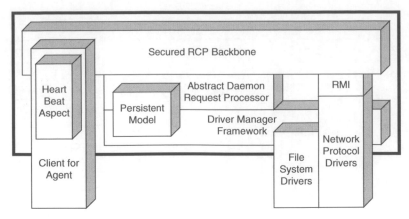

Fig. 7. GWE agent architecture.

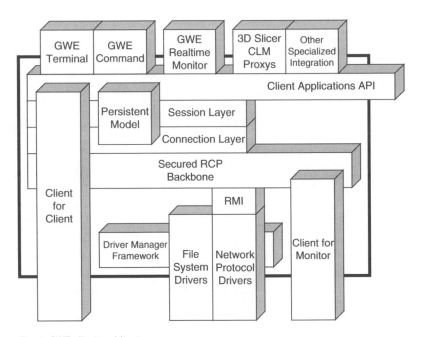

Fig. 8. GWE client architecture.

thread making information retrieval synchronous, effectively emulating real-time event notification.

- *Connection layer.* A thin module in charge of abstracting away all issues related to connecting to remote servers (connecting, losing connection, retries, reconnection, etc.).

- *Session layer.* A thin module that provides a programmatic abstraction layer to access a particular user's services "sandbox"; by transparently embedding user identification on each request required.

4. Software Development Process

GWE development is done through the following process:

1. Up to the first release of GWE (version 0.6.0.alpha) the single atomic objective was to create an infrastructure for the project and the product to allow for future organized incremental progress.

2. At the point of this release a full project management infrastructure was in place.

3. From that point forward, issues are logged into the issue management system.

4. From that list a preselected group of issues are target for resolution as the milestone for the next version.

5. When the selected issues are resolved passing the respective tests:

   a. A new release is scheduled for the following week's first business day.

   b. Its version is set as equal to the previous with the third digit incremented.

   c. A branch in the source control repository is created with a snapshot of the product's main branch and labeled with the new version number.

6. Documentation is prepared for this release (release notes, updates on obsolete and new information, etc.).

7. Regular development continues on the main branch if time left.

8. On release day:

   a. The product is checked out as a parallel project (to the currently progressing) to create the new version bundles to distribute.

   b. The project site gets updated with a news entry announcing the release of the new version and any other relevant update information.

   c. An email announcement is sent to "gwe-users" mailing list.

9. **Steps 3–7** will iterate until the product reaches a degree of overall reasonable stability and features set. At this point a new version will be released by incrementing the second digit by 2. This will signal change of phase from alpha to beta (0.8.0.beta) and from beta to general availability (1.0.0.ga).

10. **Steps 3–8** will iterate until the product reaches 1.0.0.ga. At that point a new general availability will be targeted just by incrementing the second digit from this point forward and following the same iterative process.

During this process, other project-related activities might take place in parallel (and most likely will), such as:

- GWE users support.

- Issue (new features, requests) logging.

- Issue status updates.
- Documentation updates.
- Project site updates.
- Announcement emails.
- Very minor fixes on released versions.

4.1. Open Source Community

The open source community has a great wealth of quality software in its portfolio; targeting many domain problems as well. GWE development relies heavily on many of their products and the team recognizes its pivotal importance in all aspects of creating this product. The following sections list the open source projects used to create GWE, broken down by domain problem.

4.1.1. Development Tools

The following products are the open source tools used to create GWE and to manage the project as a whole:

Product	Description
Maven	Project management tool. Oversees many aspects such as build, site generation, dependencies, bundles assembly, documentation, etc.
Maven Plug-ins	Modules used to extend "Maven" with value added features such as wiki parsers, javadoc generation, unit test execution, etc.
Eclipse	IDE (Integrated Development Environment) used to create GWE
Eclipse Plug-ins	Modules used to extend "Eclipse" with value added features such as maven integration and velocity scripts editor
Junit	Part library, part development tool. It provides a framework to create and run unit tests against the GWE source code
Cobertura	Coverage tool that calculates the effectiveness of the product unit test in terms of detecting which pieces of code are being tested and which are not
Subversion	Source control system
Apache	HTTP server used to publish project's website
Tomcat	Java web container used to run Hudson and Jira
Hudson	Automated build system
Jira	Issue tracking system. Not open source; but being used through a program which offers free license for open source projects
Mailman	Mailing list system used to communicate project announcements and support. Not open source, but free software

4.1.2. Product Libraries

The open source community has a great wealth of quality software in its portfolio; targeting many domain problems as well. GWE development relies heavily on many of their products and the team recognizes its pivotal importance in all aspects of creating.

Product	Description
Spring	Full-featured application framework. Main features exploited: dependency injection, declarative DB transaction, aspect-oriented programming
Hibernate	Object relational mapping (ORM) framework; used to abstract away the embedded database GWE uses
Derby	Java-based databases engine; used embedded as a data persistent layer
Velocity	Template engine; used in different ways across GWE
Log4J	Logging framework; used all over to create useful debugging information
Xstream	Java object to xml serialization framework; used for configuration purposes
JSch	Library to programmatically create and use SSH connections
Apache Commons	This is an umbrella for a set of open source projects targeted to deal with core type of problems. A subset of these projects is used by GWE

4.2. Project Management

GWE project is managed using Maven and using Eclipse as the IDE of choice, although any other Java IDE can be used instead without problems. Any "mavenized" project, at its core, consists of a project descriptor (pom.xml), which contains all information necessary to carry on the management tasks desired by the project team.

GWE exploits only the following management tasks out of maven:

- *Dependency management.* Declarative management of GWE dependencies and their versions.

- *Project building.* Compilation of classes using standard libraries and the dependencies declared.

- *Distributable assembly and publication.* Creation and deployment (to a predefined publication server) of bundles (zip, tar, bz2, etc.) with all the artifacts needed by a GWE product (binaries, configuration, libraries, etc.).

- *Report generation*. Generation of reports based on different metrics computed for the project such as unit test coverage, to do list, fixed defects, etc.

- *Wiki generation*. Generation of html pages based on wiki pages created.

- *Site generation and publication*. Compilation of generated reports, wiki, images, and other artifacts into a cohesive website and publication of it in a predefined Http server.

4.3. Issues Management

As an open source project, GWE has been able to take advantage of a free license of the issue management system Jira (*see* **Fig. 9**); this is an invaluable tool for dynamic projects such as GWE. This serves basically as a crosschecked list of issues with the project.

4.4. Communication

Communication with users (documentation, announcements, updates, etc.) is crucial to any project and that is why this project

Issue Navigator

Displaying issues **1** to **32** of **32** matching issues.   [Pen

Current View:
Browser | Printable | XML | Full Content (HTML | Word) | Excel (All fields | Current fields)   ☐ Bulk Change: all 32 issue(s
                                                                                                     ☐ Configure your Issue Navig

T	Key ⇧	Summary	Reporter	Pr	Status	Res	Created	Updated	Components
⊞	GWE-1	Empower value range semantics in P2L to include steps and float types	Marco Ruiz	⬆	Open	UNRESOLVED	12/Feb/08	12/Feb/08	Core: P2L
⊡	GWE-2	Fix allocations timeouts	Marco Ruiz	⬆	Open	UNRESOLVED	12/Feb/08	12/Feb/08	
⊞	GWE-3	Wildcard support on variable replacement to traverse filenames	Marco Ruiz	⬆	In Progress	UNRESOLVED	12/Feb/08	12/Feb/08	Core: P2L
⊞	GWE-4	Create a virtual file system per order to cache files reused by its jobs to improve network performance	Marco Ruiz	⬆	Open	UNRESOLVED	12/Feb/08	12/Feb/08	
⊞	GWE-5	Applications management module	Marco Ruiz	⬇	Open	UNRESOLVED	12/Feb/08	12/Feb/08	Core: New Modules
⊞	GWE-6	Make GWE JSDL complaint	Marco Ruiz	◇	Open	UNRESOLVED	12/Feb/08	12/Feb/08	Core: New Modules
⊞	GWE-7	Add support for authentication through public keys and X509 certificates	Marco Ruiz	⬆	Open	UNRESOLVED	12/Feb/08	12/Feb/08	
⊞	GWE-8	Create a module to allow descriptive result parsers	Marco Ruiz	◇	Open	UNRESOLVED	12/Feb/08	12/Feb/08	Core: New Modules
⊡	GWE-9	Clean up allocation workspaces once an allocation is disposed	Marco Ruiz	◇	Open	UNRESOLVED	12/Feb/08	12/Feb/08	
⊞	GWE-10	Run daemon requests as the user that queued the order	Marco Ruiz	⬆	Open	UNRESOLVED	12/Feb/08	12/Feb/08	
⊞	GWE-11	SRB support	Marco Ruiz	⬇	Open	UNRESOLVED	12/Feb/08	12/Feb/08	Core: Drivers Integration: LDDMN
⊠	GWE-12	Make P2L parser more robust	Marco Ruiz	⬆	In Progress	UNRESOLVED	12/Feb/08	12/Feb/08	Core: P2L
⊠	GWE-13	Save compute resources info in the DB	Marco Ruiz	⬇	Open	UNRESOLVED	12/Feb/08	12/Feb/08	
⊞	GWE-14	Alert definition and notification module	Marco Ruiz	◇	Open	UNRESOLVED	12/Feb/08	12/Feb/08	Core: New Modules
⊞	GWE-15	Command Line Module Proxys	Marco Ruiz	⬆	Open	UNRESOLVED	12/Feb/08	12/Feb/08	Integration: Slicer
⊠	GWE-16	Documentation for Slicer Integration in GWE wiki and NA-MIC wiki + tutorial videos	Marco Ruiz	⬆	Open	UNRESOLVED	12/Feb/08	12/Feb/08	Integration: Slicer
⊞	GWE-17	Create admin utilities for daemon	Marco Ruiz	⬇	Open	UNRESOLVED	12/Feb/08	12/Feb/08	

Fig. 9. GWE's JIRA snapshot.

includes two different forms of it each targeted to somehow complementary scenarios.

5. Notes

5.1. Project Site

GWE has a dedicated project website (*see* **Fig. 10**) where all relevant documentation is hosted (http://www.gridwizard.org/gwe). This website is intended for users to find any up-to-date information related to the project, from news to project and product documentation.

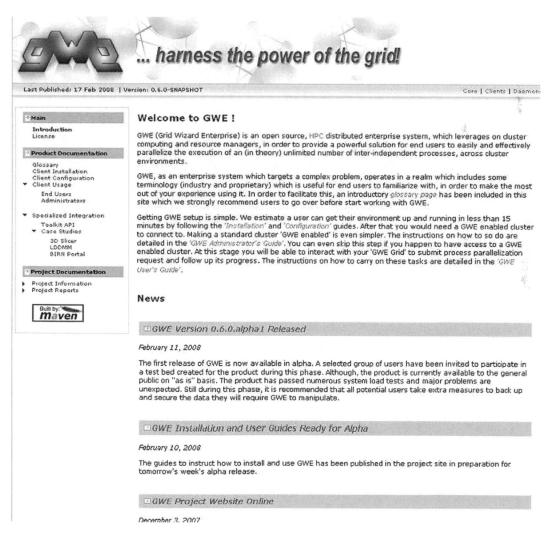

Fig. 10. GWE's project website snapshot.

5.2. Mailing Lists

GWE counts with two mailing lists intended to provide mass "on-demand" communication channels from the GWE team to GWE users and from a GWE user to the GWE team:

- *gwe-support.* Mailing list composed by the members of the GWE team. Used by GWE users to request any type of support for the product; from usage questions to report defects and request features.

- *gwe-users.* Mailing list composed by GWE users. Used by the GWE team to broadcast announcements and news to the GWE community.

References

1. J. Epema, M. Livny, R. Dantzig, X. Evers, and J. Pruyne. A worldwide flock of condors. Journal on Future Generations of Compute Systems, 12:53–65, 1996.

2. I. Foster and C. Kesselman. Globus: A Toolkit-Based Grid Architecture, pp. 259–78. Morgan Kaufmann: San Francisco, CA, 1999.

3. W. Gentzsch. Sun grid engine: Towards creating a compute power grid. Cluster Computing and the Grid, 00:35, 2001.

4. S. Pieper, B. Lorensen, W. Schroeder, and R. Kikinis. The NA-MIC Kit: ITK, VTK, pipelines, grids and 3D slicer as an open platform for the medical image computing community. Proceedings of the 3rd IEEE International Symposium on Biomedical Imaging, 1:698–701, 2006.

5. http://slicer.org/pages/Introduction, May 2008

Chapter 12

Single Sign-On in a Grid Portal

Ramil V. Manansala

Summary

Single Sign-On (SSO) is a practical requirement for software applications, which rely on distributed, networked services requiring authentication. SSO is as much a convenient feature for users as it is a security concern for application designers. The security requirement becomes critical in institutions that adhere to HIPPA regulations. In this chapter, we discuss SSO as it applies to a grid portal using remote computational resources and grid storage, which contain Personal Health Information (PHI). We cover the implementation of Public Key Infrastructure(PKI) to meet HIPPA security requirements such as authentication, confidentiality, nonrepudiation, and dataintegrity. Furthermore, we discuss the different technologies in PKI that solves these security concerns with respect to protecting research data and make SSO possible in the grid environment at the Biomedical Informatics Research Network (BIRN).

Key words: Single sign-on, PKI, BIRN, Gridsphere, MyProxy, GAMA

1. Introduction

As more software applications become more web-based, requirements for finding better solutions to securing these applications become important. Services of different varieties found on the web have increased over the years. The flux of these service-oriented sites has cultivated a need for identity providers to support Single Sign-On (SSO) mechanism to all these disparate applications on the Internet. SSO's end goal is to provide convenience as well as security. Users only need to type in their username and password once on one system and are able to use other services without having to expose their secret combination again. This "login once work anywhere feature" is convenient and avoids keeping track of multiple secret combinations for different servers.

Vadim Astakhov (ed.), *Biomedical Informatics,* Methods in Molecular Biology, vol. 569
DOI 10.1007/978-1-59745-524-4_12, © Humana Press, a part of Springer Science+Business Media, LLC 2009

It prevents possible breach of security when people start writing their information down on Post-It notes.

This chapter focuses on providing insights on how the Biomedical Informatics Research Network (BIRN) uses the Public Key Infrastructure (PKI) to adhere to the National Institute of Health's mandate of adhering to the Health Insurance Privacy and Portability Act (HIPPA) regulations. As a research institution working on sensitive patients' data, BIRN must protect the privacy of its research subjects. Data considered as Personal Health Information (PHI) should be kept private. BIRN has achieved these requirements by utilizing security infrastructure based on standards while adapting leading-edge open source technologies. Specifically, BIRN has implemented a secure SSO mechanism among distributed, networked servers and services through their portal to foster these goals. The BIRN project utilizes accepted security model to provide a secure environment for its researchers. This allows BIRN scientists to access their data, perform compute-intensive processes, analyze, and see their data using visualization tools securely. The BIRN community accesses these grid services through the browser via a grid portal.

The coming discussions focus on two encompassing topics. The first section talks about the PKI and related technologies that make it possible to implement a secure SSO mechanism in the BIRN portal. The other section provides insights to the three use-cases of BIRN portal users. The three use-cases were the driving force behind the implementation of a secure SSO mechanism at BIRN. The portal's delegation ability allows for users to login once and let the portal act on their behalf to get access to data in the grid storage, perform long-running computation in the BIRN cluster, and utilize BIRN analysis and visualization tools downloaded automatically to their local machines.

2. Motivations for Security

Why do we care about security? Why are we protecting the data and what data are we protecting? Who are we protecting them from? Questions on why, what, and who will be answered in the following section. The latter part of the chapter covers how BIRN achieves SSO and the technologies it utilizes.

2.1. HIPPA Requirements

The United States government enacted a law in 1996 to protect the Personal Health Information (PHI) of its constituents. HIPPA as it is commonly called stands for Health Insurance Privacy and Portability Act. The regulation calls for the protection of patient records by the institutions that collect and keep information. HIPPA requires the following policies implemented in the infrastructure where PHI are stored. A person needs to provide identity to gain access to records that pertain to a patient's personal health. Access to the data has to be controlled and tracked. Individuals who have the appropriate privileges can work on the data and any action on the records is logged. PHI should always be readily available as much as it is secure to authenticated entities. This implies that organizations like BIRN must protect PHI records from unauthorized entities whether they be an individual or another organization. Institutions should ensure the data integrity and confidentiality of everyone's PHI *(1)*.

2.2. Security in General

Security should be paramount to most software systems. Its implementation should cover the whole lifecycle of the application. This requirement is most crucial for software applications accessible through the web. An application should protect users from the moment they start using it, during use of services within the application, and even after having logged out of the system. Software security design requirements should focus on authentication, authorization, delegation, integrity, confidentiality, and nonrepudiation.

Authentication refers to the challenge to prove yourself as who you say you are before being allowed to gain access to a system. Authorization is granting abilities to what a user can do while authenticated in your application. Confidentiality means keeping user information private or protected from others. Nonrepudiation is the ability of your system to keep track of user actions so that users cannot deny committing an action.

The advent of PKI has provided a lot of opportunities for designers and developers to implement applications and systems that meet the issues mentioned above. The solution brought on by PKI regarding authentication, authorization, integrity, confidentiality, and nonrepudiation has made most of the web transactions we now see on the web today secure. I will talk about each of these topics in detail and use the BIRN Portal use-cases as examples for the discussion.

The BIRN Portal relies on PKI to meet HIPPA and its users' security requirements. Chief among these requirements is the ability to delegate authentication and authorization keys. The concept of delegation has become very important. Proxying as it is commonly called allows machines or processes to act on a researcher's behalf. A concrete example would be for a user named Angie to retrieve a credential from a Certificate Issuer and locally store the credential into her desktop machine. Angie starts a job submission process that requires the public-key credentials for authentication on another system. This process can run for several hours or days. Angie is not expected to wait for the jobs to complete. At the conclusion of the job submission, the system needs to authenticate to another remote host to store the resulting data of the completed job. Without the PKI mechanism this kind of operation cannot be securely implemented easily. However, with PKI, this kind of secure solution comes naturally.

3. Overview of BIRN Technology Usage

3.1. Grid Portal

Gridsphere is a Web portal designed to run portlet applications. The Gridsphere Portal Framework adheres to the portlet specification set forth in the JSR 168 document. Gridsphere is an open-source software specifically designed for grid-enabled environment. The portal framework was developed through the European Union's (EU) effort to provide tools for grid computation. The goal of the Gridsphere project is to create easy-to-use portlets that facilitate collaborations among researchers, visualize data, perform computation, and provide access to different host of services within a grid environment (2).

Besides being a portal container designed to work with grid applications, Gridsphere provides other portal features such as content management, session tracking, user administration, role creation, and group assignments. Gridsphere also supports a pluggable authentication mechanism whereby developers can implement their own authentication mechanism and use it to validate their users rather than the default Gridsphere authentication method that comes with the software.

3.2. PKI Concepts

3.2.1. PKI Using X.509

The Gridsphere portal can support multiple kinds of authentication mechanism such as MyProxy, JAAS, and MD5 to name a few. Common implementations of system authentication are shared secret and use PKI. Shared secret can take the form of username and password, One-Time Pads which contain a list of key-value pairs, and Time-Based Authentication which uses password generated by shared algorithm based on time (3). All three mechanisms work if the "shared secret" is kept private between the client and the server. PKI is a system that provides key management to protect data through encryption using both asymmetric and symmetric algorithms. PKI uses X.509 certificates but there are alternatives. Two alternatives to using X.509 with PKI are Kerberos and PGP. This chapter focuses on X.509 usage since this is the solution the BIRN project implements.

BIRN's security implementation relies on using shared secret (username and password) in tandem with PKI's GSI libraries and tools specifically, the use of X.509 certificates and the GSI tools to generate, store, and retrieve proxy credentials from an online

MyProxy server for authentication and delegation. BIRN has adopted GSI for its security solution as it has become the leading choice of other Grid efforts around the world *(4)*.

Another related PKI concept is encryption or cryptography. Encryption is a way to conceal data. A good example would be to scramble a message sent from one person to another. A third party and even the messenger would not be able to read the message if it were encoded with some type of encryption. Before PKI, most encryption relied on symmetric encoding. Messages or data were encrypted and decoded using the same key. Asymmetric encryption (also known as public key cryptography) uses one key to encrypt the data and another key to decrypt it. One of the advantages of symmetric encryption is that encryption and decryption is fast. The main disadvantage is that the data can easily be compromised since the same key to encrypt the message somehow has to be sent to the message recipient as well.

3.2.2. Cryptography Using Symmetric and Asymmetric Ciphers

We turn to discussion of the underlying mechanism of PKI before getting into the details of how it meets the security policies that HIPPA requires from institutions. PKI uses both symmetric and asymmetric ciphers to protect electronic data. So what is symmetric cryptography and why do we use it? Symmetric encryption works with only one key for scrambling data and for putting it back to its original form. The advantage of this technique is that the encryption and decryption process is very fast, as it does not involve as complex an algorithm involving factoring large prime numbers as asymmetric encryption does. Steel says that symmetric cipher is up to thousands of times faster than asymmetric cipher (p. 59). The disadvantage is that a sender who encrypted the data with the sender's secret key will provide the same exact key to the message recipient to decipher the contents of the sender's message. So to send the same message to multiple individuals, the sender would either share the same key to all the recipients or generate different keys for each one to send out. The most obvious issue with this is that anyone who has access to the sender's secret key can open all the messages, which has been encoded with the same secret key. Subsequent messages and even those messages from the past can potentially be compromised. Another issue with symmetric encoding is the need to send the key separately from the message. Sending them along together by the same means is tantamount to sending the message unencrypted. So there is an extra effort to send the key and message when using symmetric encoding. Symmetric cipher's primary use is to foster data confidentiality but lacks the other features that asymmetric ciphers provide such as nonrepudiation, data integrity, and others. Asymmetric cipher mechanism involves the same operation as the symmetric one. The cipher also encodes and decodes data with the use of keys. The main difference is that asymmetric uses one key to encrypt the data and uses another key to decrypt it.

Thus, entities within a PKI system always have both a public key and a private key. These key pairs are used in the decryption and encryption of data for authorization, authentication, concealment, and nonrepudiation. These two keys provide unique identifying information to an entity in a PKI system. The next section walks through the asymmetric encryption concepts and the security issues it solves.

3.2.3. Public–Private Key Pair

An entity such as a user, server, or other wanting use of an infrastructure which supports PKI need to have some form of credentials such as that of X.509. As mentioned above, credentials are composed of a user's private and public keys. Entities provide their public keys to other entities within a PKI system to which they wish to communicate with. At all times entities keep their private keys secure and protected. Each private key must be unique to an entity within the system; otherwise, security requirements such as data integrity, confidentiality, and nonrepudiation among others cannot be guaranteed. Data encrypted with a sender's private key and decrypted with the sender's public key guarantees that the message came from the owner of the private key and nobody else. Message encrypted with the recipient's public key assures confidentiality, message integrity, and authentication. The assurance comes from the PKI system's guarantee that only the entity that has sole possession of the private key can decode the message.

3.2.4. Authentication and Confidentiality with PKI Solution

Let us look at how authentication and confidentiality is achieved within a PKI. Data can be secured from unintended recipients of a message through encryption using the recipient's public key. Encryption of data using a sender's private key provides authentication. If Mark wants to send Angie a secured message, Mark requests from Angie his public key. Upon receipt, Mark encrypts the message with Angie's public key. This secures the message through transit on the network because only Angie has the private key that can decrypt the encrypted message. Even Mark could not reconstruct his own message to Angie after it has been encrypted with Angie's public key. Both Angie and Mark can rest assured that the messages exchanged between them are secured. In the same token, both Angie and Mark have authenticated to each other since messages decrypted with Angie's public key could have only come from Angie who has the private key that encrypted the message for Mark. This applies to Mark's situation as well. This is really the power of asymmetric ciphers. The drawback of using an asymmetric cipher this way is its inefficiency of encrypting and decrypting large data. Asymmetric ciphers are thousands of times slower than symmetric ciphers *(1)*. The elegant solution is combining the symmetric and asymmetric encryption techniques in securing the message. The following scenario illustrates the use of both symmetric and asymmetric ciphers to encrypt a message. Mark generates a random temporary symmetric key and uses that to encrypt the actual data using symmetric cipher. Mark then creates a new document, which contains the newly generated temporary key and encrypts the document with Angie's public key using asymmetric cipher. Mark appends this new encrypted document with the original encoded message. He sends these combined data to Angie. Angie uses her private key to decrypt the smaller message containing the symmetric key. Angie then uses the symmetric key to decode the much larger symmetrically encrypted data back to its original state. This technique is advantageous because of two things, speed and convenience. The encoding and decoding of messages is faster because the larger message is encrypted with the much faster symmetric algorithm while the smaller document containing the key is encrypted with the much more secure asymmetric cipher. Using a combination of the two ciphers this way is convenient due to the fact that both message data and key can be sent together in one message.

3.2.5. Data Integrity and Nonrepudiation with PKI

Using public keys in ensuring data integrity puts a different emphasis on the exact same technology illustrated in the previous section. Usage of digital signature within a PKI environment explains the concept of data integrity. Digital signature is extra electronic information added to digital data, which can assert authenticity and integrity to the data. Digital signature is created to protect documents from corruption or tampering during transmission over the network. One way PKI implements digital signatures is to use the reverse of how the usual asymmetric encryption works. In this scenario, Mark encrypts the message with his private key to send Angie a secure message. Angie uses Mark's public key to decrypt the document to read it. This assures Angie that the document must have come from Mark since the only ways he was able to decrypt it using Mark's public key is that it must have been encrypted with Mark's private key. Moreover, Mark has the sole possession of the private key associated with the public key that opened the message. But the reason for encrypting data in this scenario is moot. Mark's public key is available to everyone as it is public. So encrypting the data before transmission does not really serve any purpose because everyone who possesses Mark's public key can open his encrypted document. The application of another PKI technology solves the problem of data integrity. There is an added requirement that the document sent across the network is not tampered with during transit. Let us take another example of a mortgage lender named John who needs to let a homeowner named Joy sign a Mortgage Agreement. John sends an electronic text copy of the agreement to Joy and Joy signs the agreement electronically by typing her name on it. Joy computes a hash value from the signed agreement and pastes this hash to a newly created electronic text document.

Joy encrypts the new smaller document containing the hash with her private key. She then attaches this new encrypted hash-containing document to the signed Mortgage Agreement document. Joy may or may not choose to encrypt the original electronic agreement she signed with her name before transmitting it back to John over the network because of the fact that everyone who has Joy's public key can easily decrypt it anyway. But by attaching an encrypted document with a hash value, Joy has effectively created a digital signature that will give John a way to check the integrity of the returned original mortgage agreement, which Joy has signed with her name. Joy sends back the document across the wire confident that any alteration in this document in any way will generate a different hash number. John uses Joy's public key to open the encrypted digital signature that contains the message digest (a.k.a. hash) value. John in turn calculates a hash value on the original signed agreement and compares this value to the one sent and encrypted by Joy. If they match, then John can be assured of two things: *(1)* that the document is complete and has not been tampered with during transmission over the network and *(2)* that Joy could not somehow claim that she was not the one who signed the document since John was able to decrypt it using Joy's public key. Assuming that Joy has the sole possession of her private key, there is no one else who could have signed the hash-containing message other than Joy. The first assurance fosters data integrity while the second guarantee nonrepudiation. All these discussions of PKI concepts relate to how BIRN has achieved to provide a Single Sign-On authentication and delegation through a grid portal to access computational services and grid storage. Let us turn our discussion to the use of Grid Security Infrastructure (GSI), X.509 Certificates, and an online proxy credential repository at BIRN that enable cluster computation and grid storage.

3.3. Grid Security Infrastructure (GSI)

GSI is built on concepts brought about by public key cryptography. The infrastructure fosters authentication and secure communication on an open network *(5)*. GSI uses X.509 Certificates, Secure Socket Layer (SSL), digital signatures, mutual authentication, delegation, and SSO. Its primary goal is to provide secure communication between entities within a grid environment. It also tries to promote ability to authenticate entities across federated organizations and, **finally**, to promote SSO for grid users performing computation in the grid *(6)*.

3.3.1. X.509 Certificates

Digital Certificates also known as X.509 Certificates are documents that contain information about the owner such as Name, Organization, Country, and State Information *(1)*. What makes this document important in the PKI environment is that other digital information is included in the document which vouches for the validity of the information regarding the owner of the certificate. A thirdparty has digitally signed your certificate to signify that you are indeed who you say you are. The burden of proof is left to the third-party entities which for the most part are publicly recognized Certificate Authorities (CA). CAs such as Verisign or Thawte are considered End Entity Certificates because their certificates are self-signed. They are publicly trusted companies that can vouch for other entities who request certification from them. Getting a digital certificate is as simple as creating a Certificate Signing Request (CSR) and sending it to the Certificate Authority like Verisign, Inc. Verisign verifies information you provide in the CSR by phone call or other means before providing you a Digital Certificate with their company signature in it.

3.3.2. Secure Sockets Layer (SSL)

In 1996, Netscape created SSL, an end-to-end communication protocol designed to secure not only HTTP between the web client browser and the web server but any data transmitted over reliable protocol such as TCP/IP *(1)*. The protocol uses PKI mechanism to authenticate the client and the server to each other as well as securing the data transmitted between them. During SSL connection there are four types of messages exchanged between the communicating parties. The message exchange start with an SSL handshake. The handshake involves authentication and choice of encryption. The security parameter exchanged between parties is also known as CipherSuite. Alert comes next followed by change of Cipher specification request. Alerts indicate whether there are fatal errors or warnings, while ChangeCipherSpec determines subsequent communication protocol agreed upon in the initial handshake. Finally, the parties exchange actual data securely.

SSL is sometimes used interchangeably with Transport Layer Security (TLS). However, according to Steel et al. SSL and TLS are not the same (p. 67). TLS was authored in 1999 by Certicom to create an IETF standard protocol. The protocol was designed out of SSL 3.0. SSL and TLS are not compatible but they are similar enough in design that TLS 1.0 can come back down to SSL 3.0. TLS sometimes is referred to as SSL 3.1.

SSL is used within the GSI environment for authentication, data integrity, and confidentiality *(7)*. Authentication is when an entity can verify another's identity. Data integrity means that data passes from one entity to another without being tampered with. Confidentiality refers to a concept in PKI when messages are protected or hidden from unintended recipients.

3.3.3. Proxy Credentials

GSI is built on PKI. In a PKI environment entities identify themselves with Grid credentials. These credentials are comprised of private key and a certificate. The certificate essentially is identifying information that specifically ties an entity to a unique name in the grid known as Distinguished Name (DN). A DN is globally unique. The DN string maps to a particular user in the grid. In Unix or Linux machines the collection of user DN strings is stored in a grid-mapfile document. Each DN is listed with a username. In GSI authentication, a certificate represents the link between an entity's private key and its DN. User validation is done through the use of Grid certificates. Users and services possess unique credential for identification and authentication purposes within the Grid *(6)*. Grid credentials include a subject which serves as the entity's unique name in the grid, the public key of the entity, the identity of the CA that signed the certificate, and the digital signature of the CA. Third-party CA affirms this correlation between the DN and private key. The CA basically vouches for the entity's claim that he is who he says he is and that the DN is his identifying name in the Grid *(7)*.

Entities in a PKI system should protect their private keys. A compromise of an entity's private key can lead to impersonation of the original owner's identity within the system. Therefore, the protection of these private keys is very important. Usually, enforcing stricter access permissions on the file or file encryption protects the private keys. Another way of protecting private keys is by setting expiration on their validity. Such time limitation makes private keys useless when they expire. The private keys can no longer be used for validating the holder of the certificate. This expiration lifetime is on the order of years *(7)*.

The above constraints for securing the private keys can affect system usability. The security measures affect individual users or services authenticating to multiple hosts in a grid environment. The choice to implement one security framework over another could be based on usability issue as well as security considerations. Novotny explains that asking users to enter their passphrase multiple times to login to several hosts in the grid is not only inconvenient but also dangerous (p. 2). The danger is when long-term credentials are exposed when they are decrypted during use. An alternative proposed in the paper to either retain the user's passphrase or the long-term credential helps with the user experience but poses a greater risk to the user's private key.

The use of proxy credentials is the better solution. GSI offers an ability to authenticate a user with a proxy certificate. A proxy certificate is a short-term binding to a user's unique DN and a new private key. Proxy credentials are created by the user and signed with his long-term credentials. Proxy credentials are not encrypted to avoid having to require the users to provide passwords to decode them for use. Because of this transparency, proxy credentials have a shorter lifetime, which is on the order of hours, if not days.

3.3.4. Delegation

Another concept in GSI system is delegation. One entity in the Grid is able to provide another its credentials so that the latter can act on the former's behalf. The power of delegation in the grid computing environment is that a user can submit a long-running computational process from one machine and be able to leave the software application running overnight. The submitted process will perform subprocesses or subtasks until they complete. These tasks could range from computation on one or several host machines within the cluster to data storage in the grid. These tasks are performed without having to ask the user to be present and enter passwords to authenticate to all the other hosts used in the computation and storage. That is the power of delegation. Delegation is only possible with the use of proxy credentials.

3.4. MyProxy Server

The MyProxy Credential Repository System was born out of the need to provide an online host where users can securely store their long-term GSI credentials, generate a proxy, and retrieve the proxy for use in authenticating to GSI-enabled hosts and services. The MyProxy Credential Repository server consists of proxy delegation and retrieval utilities as well as storage for the credentials.

Usage of the MyProxy server starts with the myproxy-init client tool along with the user's long-term credentials in his filesystem. This process delegates to the MyProxy server the user's proxy credentials along with the username and passphrase combination and other limitations for later retrieval of the user's proxy (7). To retrieve a proxy, users will use myproxy-get-delegation client tool. To invalidate the user proxy credential another client tool is invoked, myproxy-destroy.

3.5. Grid Account Management Architecture (GAMA)

BIRN uses a MyProxy server to authenticate users into their system via the portal. The portal serves as the gateway to all the different services that BIRN creates for its community. These services are independent of the portal server and require authentication to use. MyProxy credentials facilitate a mechanism that enables BIRN to integrate all these different services to allow users access to their system.

BIRN uses the Grid Account Management Architecture (GAMA) to integrate both key management of X.509 Certificates and delegation of MyProxy proxy credentials from a single credential repository. The main purpose of GAMA is to alleviate users' burden of managing their long-time GSI credentials. In a nutshell, GAMA provides web services for the creation and management of Grid accounts for a JSR-168 portal (8). Users who wish to participate in a GSI-enabled environment do not have to go through creating their public and private keys to start delegating their proxy credentials in the MyProxy repository. In addition, users do not have to worry about safeguarding their private keys that would otherwise have been in their local filesystem. The GAMA software stack includes web services and portlets among others which provide a wrapper for automatic PKI public and private key generation and proxy delegation functionalities.

The GAMA server is locked-down from any other access except the portal. The only connections that the GAMA server passes through are the ones coming from the BIRN portal. GAMA provides HTTP over SSL configuration options for further securing the system as well as HTTP for quick and easy deployment.

GAMA has a two-step process for both account creation and password renewal. GAMA has a separate web service call to create the End Entity Certificate EEC credentials using the NAREGI Certificate Authority (CA). Another web service call creates the MyProxy account of the user. Password changes involve the deletion and re-creation of both the EEC key pairs and their storage to the MyProxy repository.

3.6. Storage Resource Broker (SRB)

The SRB is a data management software that provides hierarchical logical organization of folders and files in the grid (9). SRB was developed at SDSC as a joint venture between General Atomics and the National Institutes of Health. The joint effort has produced a middleware that provides the ability to conglomerate distributed and disparate data storage that includes disks, databases, filesystems, and tapes over the network. SRB has a relational database server that acts as the index for how and where data are stored physically. This master server is called the MCAT or Meta Catalogue server as it holds the information how these data and users are grouped logically.

As of this writing, SRB supports only two kinds of authentication mechanism. The simple password-based authentication has become obsolete (9). Currently, GSI and Encrypt1 are supported. BIRN uses GSI authentication mechanism to connect users of the portal to their SRB user accounts.

3.7. Condor

Condor is a job scheduling and management software designed to help engineers and scientists to harness the power of a collection of distributed, networked computers. The Condor project's goal is to increase the computing throughput of their users, specifically engineers and scientists (10). GWiz prepares the environment including the staging of files from SRB in the BIRN cluster head node for all the jobs before handing it off to Condor. Condor takes care of assigning to each of the 24 BIRN cluster compute nodes which job subtasks to run.

3.8. Grid Wizard (GWiz)

GWiz allows users to submit jobs on the command-line and through a grid portal. Jones (11) states that GWiz can simultaneously execute tens of thousands of tasks in a cluster environment. The application uses Condor for its job scheduling mechanism but it also supports others such as Sun Grid Engine. We discuss Condor below as it is the one we are using at BIRN. GWiz is an open-source software developed primarily in Java. It was developed at BIRN through funding from NA-MIC and BIRN.

3.9. Java Network Launch Protocol (JNLP)

Java Specification Request (JSR 56), also known as Java Network Launch Protocol (JNLP), is a specification of Application Programming Interface (API) for enabling Java applications to work in the web environment. JSR 56 API solves the issue of distribution of Java applications. The specification promotes the delivery of applications to the user via a web server and launched from a web browser. The document also covers the re-delivery of updated software to various target platforms such as desktop, server, personal computer, and others. The proposed specification extends the applet model to support web deployment and run rich client applications at the same time. It uses the applet's sandbox model for security. The difference is that while the applet lifecycle depends on a web browser, the Java application launched via JNLP runs independent of the browser *(12)*. Once launched, the web page that served the application can go away without effect on the running Java Web Start program. The windows and processes associated with it are agnostic of the local machine's operating system and the web server that delivered it. The program only cares about the Java Runtime Environment or the "sandbox" it is going to operate in. The framework will make it easier for developers to deliver tools and content while it facilitates ease of use for their users when launching applications with a click of an HTML document link.

3.10. Java Web Start

Java Web Start is a mechanism to deliver a Java application to a user's desktop through a web server *(13)*. Rich client application can now be part of a web site to provide functionalities difficult if not impossible from a pure browser-based application. This is totally different from Java applets which run within the context of a web page. A Java Web Start application is downloaded to the user's machine and runs in a sandbox within the user's local machine. This sandbox is a contained environment of the user's Java Runtime Environment (JRE). So, the application runs outside the context of the browser but within the confines of the JRE. The advantage is that the Java program can have underlying information that came from the server and application that served it as well as having access to the local machine's filesystem.

3.11. ImageJ

ImageJ is an image processor for a variety of formats. It can analyze, edit, and save TIFF, GIF, DICOM files, FITS, JPEG, as well as BMP. The program is multithreaded which allows compute-intensive operations to be executed in parallel. Features include the ability to calculate area by pixel value statistics *(14)*, measurement of angles and distances, and display of density histograms. In addition, ImageJ supports standard image processing capabilities such as sharpening, smoothing, and contrast adjustments to mention a few. Furthermore, the program provides geometric transformations features which include but are not limited to rotation and scaling.

ImageJ has a pluggable architecture. Developers can get the source code and develop plugins to extend the features of ImageJ. The plugins help users overcome issues they encounter with their image processing and analysis process.

ImageJ is freely available and is part of the public domain. The application has two versions: an applet and a stand-alone Java application. The applet runs on a browser while the downloadable application will run on any computer that has a JRE 1.4 or later installed on their system.

4. BIRN Use Cases

4.1. Accessing Grid Storage

4.1.1. Motivation

Traditional web portals which do not have support for grid computing have difficulty providing SSO capability with the security as strong as PKI-based encryption. On the one hand, it is possible to provide users secure SSO capability by using X.509-based proxy certificates. Users wanting to take advantage of services in the grid using GSI authentication need only to retrieve their proxy credential from the online MyProxy server. However, they need to first create their own X.509-based public and private key and store them in the MyProxy server. The MyProxy server requests a username and password combination for storage as these information are needed for subsequent retrieval of a Proxy Certificate. Once a user is in possession of their proxy credentials they can then use the proxy to access large disk storage resources on the grid. On the other hand, BIRN has implemented a user account management and authentication system two-steps beyond this solution implemented on other academic and research portals. By leveraging the GAMA web services wrapper to the MyProxy server functionality, BIRN has delegated the End Entity Certificate (EEC) creation of every user as well as the storage of these keys into the MyProxy server itself. The MyProxy server in GAMA has a secure online X.509 credential repository that has proxy delegation capability. In addition, the web services in both systems include the ability to retrieve the user's proxy credentials by providing a username and password combination.

4.1.2. SSO Process

Putting it all together, here is how the BIRN portal achieves SSO among three separate servers. This use-case scenario involves the grid portal, the identity provider, and the grid storage server (refer to **Fig. 1**). The SSO process starts with the user's browser talking to the portal server over SSL. The server responds to the client by presenting it with the portal's X.509 digital certificate (refer to **Figs. 2** and **3**). The client browser can alert with error

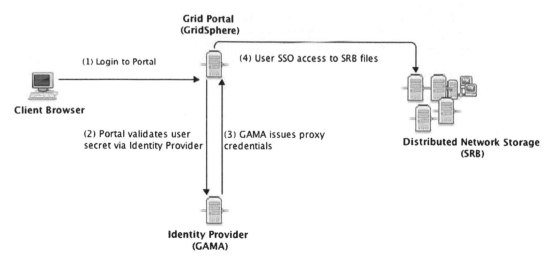

Fig. 1. Portal SSO mechanism to grid storage.

Fig. 2. (**a**) SSL handshake between client browser and BIRN portal. (**b**) BIRN portal login.

or warning; otherwise, the browser proceeds with the handshake and generates a random value and encrypts this with the grid portal's public key included with the digital certificate. The client sends a ChangeCipherSuite message along with the encrypted random value. The server decrypts the random value with its pri-

Fig. 3. Storage resource broker access from the grid portal.

vate key and responds with a ChangeCiphereSuite. The client browser and the portal now share a shared "secret" in the random value. Steel says that this "shared" secret value is then used to create a master secret from which keys are derived used to encrypt actual data transmitted after the SSL handshake (p. 70).

After the initial handshake, communications are passed securely between the client and the portal thus allowing for the portal to use its own credential to authenticate itself to the MyProxy server within GAMA and provide the portal user's information (username and password combination) to the credential repository. This step accesses a user's long-term credential in GAMA and creates a short-term proxy GSI credential. The portal subsequently uses this proxy to authenticate to the grid storage account (SRB) on behalf of the user.

4.2. Submitting Remote Jobs

The SSO process for submitting remote jobs start out in the same way. BIRN portal user signs on to the portal (**Fig. 4**). The portal retrieves the user's X.509 proxy credentials from GAMA. The portal uses the proxy credentials to authenticate the user automatically to the SRB using GSI (**Fig. 3**). The user is redirected to his SRB home folder and locates the folder where the image data are located. The user selects an image atlas to run against the other target images in the directory and launches the Large Deformation Diffeomorphic Metric Mapping (LDDMM) *(15)* job (**Fig. 5**). LDDMM is used to compare large, complex brain structures and detect variances using the computational grid *(16)* at BIRN and TeraGrid. The images along with the user's X.509 certificates are

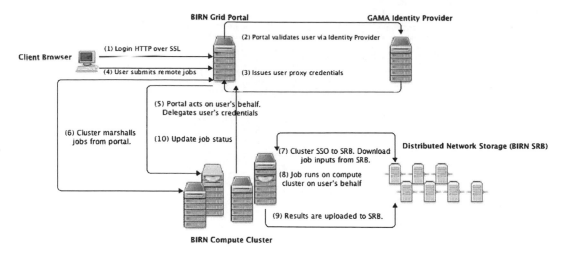

Fig. 4. Remote jobs submission from the grid portal.

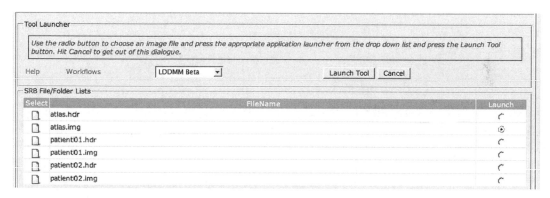

Fig. 5. Submitting the job from the grid portal interface.

then downloaded from the portal and staged in the BIRN cluster for preprocessing. Each job's subtasks are submitted to Condor. Condor queues each task and runs them on available nodes in the cluster. Users can monitor the status of the submitted jobs from the portal user interface (**Fig. 6**). When the LDDMM jobs complete, their results are copied to the SRB (refer to **Fig. 7**).

4.3. Analysis and Visualization

The analysis and visualization SSO process works a little differently from the previous two use cases. In this scenario the application, security credentials, and the data are sent across the wire from the server into the user's desktop via the browser. It starts with the user sending the username and password information to the portal via HTTP over SSL. The portal then accesses the GAMA authentication server to retrieve the user's MyProxy credential using the user secret (*see* **Figs. 8** and **9**). The proxy certificate retrieved by the portal is stored in the Gridsphere

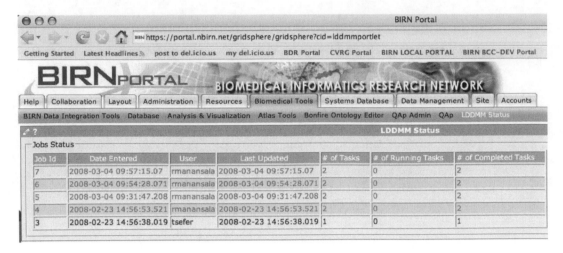

Fig. 6. Monitoring the job from the portal interface.

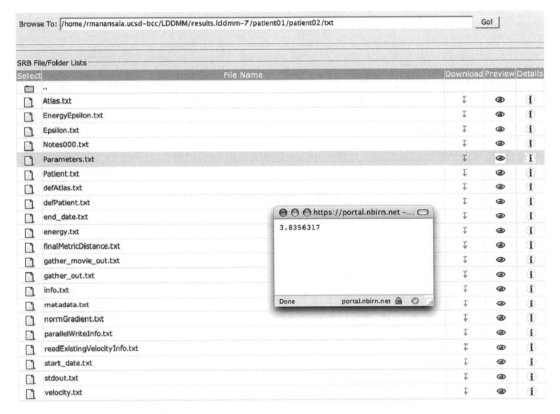

Fig. 7. Reviewing the job output in grid storage.

Portal. The user proxy persists temporarily in memory within the Portal container. The user clicks on the ImageJ link to launch the application, and the portal sets the users connection parameters to the grid storage along with the user's proxy certificate into the JNLP file (*see* **Figs. 10** and **11**). ImageJ is launched

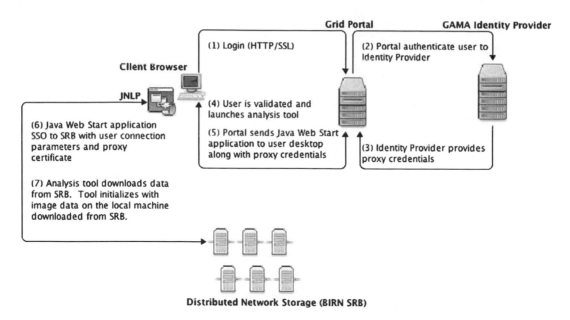

Fig. 8. Portal SSO with analysis and visualization tools.

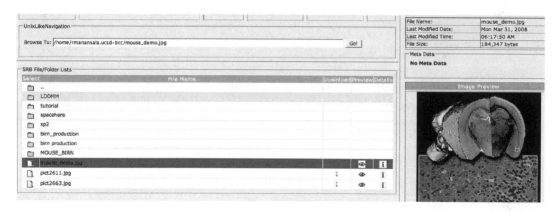

Fig. 9. Selecting image data to analyze from the grid storage.

from the user's computer along with a helper application, which sets the user connection parameters and proxy credentials. The application connects automatically to the data grid storage via GSI authentication mechanism. The user is authenticated to the SRB without further username and password inputs from the user. The files selected during the launch are downloaded automatically to the user's local machine. The ImageJ application launches subsequently and opens the downloaded file from the grid (*see* **Fig. 12**). The SSO process is complete in this scenario. The portal user is ready to work with the ImageJ application running on his local machine with his image files already loaded in the application.

```
<property name="srb.host" value="bcc-gpop.nbirn.net"/>
<property name="srb.port" value="5825"/>
<property name="user.name" value="rmanansala"/>
<property name="home.dir" value="/home/rmanansala.ucsd-bcc"/>
<property name="srb.domain" value="ucsd-bcc"/>
<property name="srb.resource" value="ucsd-bcc-nas"/>
<property name="proxy.str" value="-----BEGIN CERTIFICATE-----
```
```
MIICJTCCAY6gAwIBAgICAg4wDQYJKoZIhvcNAQEEBQAwaDELMAkGA1UEBhMCVVMx
DTALBgNVBAoTBEJJUk4xEDAOBgNVBAsTB0JJUk4tQ0MxEzARBgNVBAMTCnJtYW5h
bnNhbGExIzAhBgkqhkiG9w0BCQEWFHJhbWIsQG5jbWlyLnVjc2QuZWR1MB4XDTA5
MDYxNTE3MjQ0NVoXDTA5MDYxNjA1Mjk0NVoweDELMAkGA1UEBhMCVVMxDTALBgNV
BAoTBEJJUk4xEDAOBgNVBAsTB0JJUk4tQ0MxEzARBgNVBAMTCnJtYW5hbnNhbGEx
IzAhBgkqhkiG9w0BCQEWFHJhbWIsQG5jbWlyLnVjc2QuZWR1MQ4wDAYDVQQDEwVw
cm94eTBcMA0GCSqGSIb3DQEBAQUAA0sAMEgCQQC5yAwg61HWrMK1rCNNlnslcGjr
4XkbCBSwOolCgzX/U4J6jxnLGP+4jRIvvmOjLrJmxHU8dfQjXKEGT/qez/mtAgMB
AAGjEjAQMA4GA1UdDwEB/wQEAwIEsDANBgkqhkiG9w0BAQQFAAOBgQBETkxpAO8x
OUaaSIXdw/r+o7i3nCy+DT++CfoV5IAVYxgPcIELl+Ye0qksoFYWt/l5yssGukuj
8wJ8XHuDeBafVbL/mJqCQmn/qsfqNBSkJjQjuSAUAkxycbFTVR1kclp59y20jz8h
c1KyxAX4hPJ++li90Gbxsu+Bdotomi4f8Q==
-----END CERTIFICATE-----
-----BEGIN RSA PRIVATE KEY-----
MIIBOQIBAAJBALnIDCDrUdaswrWsl02WewhwaOvheRsIFLA6iUKDNf9TgnqPGcsY
/7iNEi++Y6MusmbEdTx19CNcoQZP+p7P+a0CAwEAAQJAScWOS2QmU0C28L0yJ5Uh
tfAtOb2HAZzr2i7kYm0H/CLXqhO5LHUC15WGsFvnGQkcggnjiQaNIL/A/HoFl1Sw
AQIhAObhw+lYDnl9XmTwEegTomTY2gU3oZFtwBZTiiUgnoaBAiEAzf4zTRpjQCfq
75xaVCny3mdexmYolejegwXs7QKJ1S0CIDXEUbevXn3gKMFc3u6q0+CkV3FrmSai
5l7MqZkn1VMBAiBDzdyNEJaC06DdZhXFhlIPl69qls22+7PiZNzr1t4PsQIgKUAb
xTuEEiwvdy2cni9g1vi76ckdG3CVNXdNyvTfpQA=
-----END RSA PRIVATE KEY-----
-----BEGIN CERTIFICATE-----
MIICmDCCAgGgAwIBAgICAg4wDQYJKoZIhvcNAQEFBQAwXDELMAkGA1UEBhMCVVMx
DTALBgNVBAoTBEJJUk4xEDAOBgNVBAsTB0JJUk4tQ0MxLDAqBgNVBAMTI0JJUk4g
UHJvZHVjdGlvbiBDZXJ0aWZpY2F0ZSBNYW5hZ2VyMB4XDTA4MDQwMzE2ENTg1N1oX
DTExMDQwMzE6ENTg1N1owaDELMAkGA1UEBhMCVVMxDTALBgNVBAoTBEJJUk4xEDAO
BgNVBAsTB0JJUk4tQ0MxEzARBgNVBAMTCnJtYW5hbnNhbGExIzAhBgkqhkiG9w0B
CQEWFHJhbWIsQG5jbWlyLnVjc2QuZWR1MIGfMA0GCSqGSIb3DQEBAQUAA4GNADCB
iQKBgQC/3K6KRkjPRATmwdnITlsHt7ArmPtelV29u6ZQGxFuT2ICJ/v6RTVQFWm0
LDiJHza/t+cTiojNSA4KDRTPmLCUs/125nqHyeBCRJQwWTiVRKBznJyB8gxUADC0
0AH3ct8CUEEl1gC7cywVZqi0qQxrHw83eGFaBK6aTPykUwsLywIDAQABo10wWzAM
BgNVHRMBAf8EAjAAMAsGA1UdDwQEAwIE8DAfBgNVHSMEGDAWgBR/58paWiXstU/t
HyVZJ33Wh1XqGTAdBgNVHQ4EFgQUslmosnlUEVr//oRADY6LHUOGFDgwDQYJKoZI
hvcNAQEFBQADgYEA25EK8Wi6nTPO4/0nwFinyXtqKl5N7Qf9xaCljbau3bEFrOag
SfBpVvILsYuxUjy5x3tvCuQu1YepfvalpvrQrAh92eLy3+daRo00JHcLIUqpqc9d
AIq1RhKYEidH4364YkrVuRHZdtZsiZxbxePaBQ8dGDiA6A3qZiCT8pTtMYc=
-----END CERTIFICATE-----
```
```
"/>
<property name="app.name" value="ImageJ"/>
<property name="app.jar" value="imagej.jar"/>
<property name="data.type" value="simple"/>
<property name="app.input" value="/home/rmanansala.ucsd-bcc"/>
<property name="down.type" value="1level"/>
<property name="cmd.opt" value=""/>
<property name="cmd.args" value="mouse_demo.jpg"/>
```

Fig. 10. Using username and proxy for SSO with GSI host.

5. Summary

The chapter covers the implementation of secure Single Sign-On using the Public Key Infrastructure in a grid environment. The sections discuss the integration of the Grid Security Infrastructure model and the online MyProxy credential repository to implement the SSO requirement of the BIRN portal using X.509

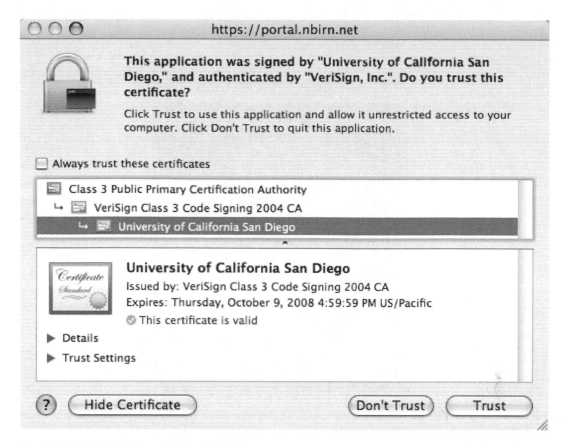

Fig. 11. BIRN digital certificate signed by CA (Verisign).

proxy certificates. It shows how BIRN utilizes open-source software to help researchers have easy access to their data in the grid, perform intensive computational analysis on the same images in the BIRN cluster, and perform visualization and analysis on the results all from one web browser session.

6. Notes

6.1. Meeting HIPPA Security Requirements

Institutions try to create a secure environment for their stakeholders. They utilize the most current tools available. They design and implement technologies that solve their problem domain and suit their infrastructure. As a leading biomedical informatics institution, BIRN has strived for the same goals. BIRN has been very successful in meeting their goals in providing a secure grid infrastructure for its research community.

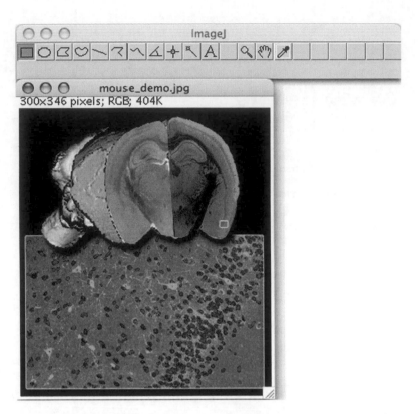

Fig. 12. Analysis and visualization tool running on local machine with image data from the grid storage.

BIRN researchers have a secure infrastructure where they can perform their tasks and collaborate with colleagues. With the help of open-source technologies, BIRN has adhered to HIPPA requirements of protecting patients' data. However, like most organizations, BIRN acknowledges places within its infrastructure that can be strengthened. The following system and services are specifically identified.

6.2. Avoiding SQL Injection

BIRN has to pay attention to user interfaces (UIs) that are available through the public. These front-facing UIs can be the gateway to attacks into the system. Web Forms such as those used for logins, registration, forums, blogs, and other similar UIs receiving user inputs that are stored in a database can be good candidates for SQL injection attacks. Although filtering for single or double quotation marks can safeguard you from malicious attacks, that might not be enough as there are a lot of ways of representing these quotation marks that get past your filters. BIRN will investigate the use of substitution parameters in its SQL queries within Hibernate to avert a higher percentage of SQL injection attacks.

6.3. Prevent Script Attacks

Blogs and forums are the most viable avenues for scripting attacks as well. The malicious scripts are inputted in blogs and forums and stored in the database. Subsequent render of the pages containing entries with embedded scripts can execute whatever attack may be located in the web page. It could be as simple as redirecting your page to another website or it could be something that forwards your current cookie information to another user, which can compromise your current session.

6.4. Investigate Securing Connection to Portal Database

BIRN will investigate the current implementation of how databases can be further secured. Besides limiting database connections by IP filtering, BIRN will look at how we can identify the tables that are accessed from these different machines and create appropriate database accounts with appropriate privileges to certain tables. Another measure is to impose periodic change of username and password to the database as well as having different username and password for development and production environments.

6.5. Securing Web Services (Putting a Lot of XML to Break SAX Parser of Axis)

Another issue BIRN needs to investigate is protecting its web services from Denial of Service (DoS) attacks. One such DoS attack can come in the form of a large payload in the XML request sent to BIRN web services causing XML parsers to break or slow down to a halt. BIRN can combat a DoS by putting some kind of filter on requests that come continuously for a period of time from the same origin *(1)*. These filters can be implemented in the applications themselves or in the system. Programs can be made to check for the payload size (XML length) or routers can be installed to filter the packet requests. Each solution has advantages and disadvantages that must be weighed accordingly. Implementing either or both is better than not having protection against an attack.

6.6. Java Web Start (JWS) Vulnerabilities in a Shared or Multi-User Environment

Finally, securing the temporary JNLP session within a shared system should be addressed. The JWS application downloaded to a shared desktop can be relaunched to access files in the SRB as long as the temporary proxy credentials are still valid. The BIRN has currently set the proxy lifetime to be valid up to 12 hours. If the shared system is not configured to have privacy settings for individual user and files then, there is the possibility of other users gaining access to another's SRB account via the downloaded Web Start application. This is an issue neither the web browser nor the Java Web Start try to resolve *(17)*. BIRN will need to both educate the portal users about this vulnerability as well as implement post-processing scripts to delete the JNLP file or invalidate the 12-hour user session after closing the application.

References

1. Steel, C., Nagappan, R., & Lai, R. (2005).*Core Security Patterns Best Practices and Strategies for J2EE, Web Services, and Identity Management.* Upper Saddle River, NJ: Pearson Education Inc.

2. *Grid Portals Introduction (2004).* Retrieved November 17, 2008, from http://www.gridlab. org/WorkPackages/wp-4/

3. *Authentication and Identification Methods* (2008). Retrieved February 10, 2008, from http://developer.apple.com/documenta-tion/Security/Conceptual/Security_Over-view/Concepts/chapter_3_section_3.html

4. Tuecke, S., Welch, V., Engert, D., Pearlman, L., & Thompson, M. (2004). Internet X.509 Public Key Infrastructure (PKI) Proxy Certifi-cate Profile, *Internet Engineering Task Force Request For Comments 3820,* IETF Website, http://www.ietf.org/rfc/rfc3820.txt

5. *Security Documentation* (2008). Retrieved April 5, 2008, from http://www-unix.glo-bus. org/toolkit/docs/3.2/security.html

6. *Overview of the Grid Security Infrastruc-ture* (2008). Retrieved April 12, 2008, from http://www-unix.globus.org/toolkit/docs/3.2/gsi/key/index.html

7. Novotny, J., Tuecke, S., & Welch, V. (2001). An Online Credential Repository for the Grid; MyProxy, *Proceedings of the Tenth International Symposium on High Performance Distributed Computing (HPDC-10),* IEEE Press, August 2001. http://myproxy.ncsa.uiuc.edu/

8. *Grid Account Management Architecture* (2008). Retrieved December 2, 2007, from http://grid-devel.sdsc.edu/gd/node/172

9. *About SRB* (2008). Retrieved March 2, 2008, from http://www.sdsc.edu/srb/index.php/Main_Page

10. *Condor High Throughput Computing* (2008). Retrieved February 20, 2008, from http://www.cs.wisc.edu/condor/

11. Jones, N. (2007). *GridWizard is... .* Retrieved December 5, 2007, from http://forums.grid-sphere.org/index.shtml

12. *JSR 56: Java Network Launching Protocol and API* (n.d.). Retrieved December 30, 2007, from http://jcp.org/en/jsr/detail?id=56

13. Zukowski, J. (2002). *Deploying Software with JNLP and Java Web Start.* Retrieved Decem-ber 3, 2007, from http://java.sun.com/developer/technicalArticles/Programming/jnlp/

14. *ImageJ Image Processing and Analysis in Java* (n.d.). Retrieved January 29, 2008, from http://rsb.info.nih.gov/ij/

15. *Software @ CIS: lddmm-volume* (2007). Retrieved March 18, 2008, from http://cis. jhu.edu/software/lddmm/

16. *BRAIN Morphometry BIRN – Atlas Tools* (n.d.). Retrieved March 18, 2008, from http://www.nbirn.net/research/morphom-etry/analysis_tools.shtm

17. Basney, J., Fleury, T., & Welch, V. (2006). Single Sign-On for Java Web Start Applica-tions Using MyProxy, *Proceedings of the ACM Workshop on Secure Web Services (associated with the 13th ACM Conference on Computer and Communications Security),* November 3, 2006. http://myproxy.ncsa.uiuc.edu/

INDEX